MIRACLES

Books by C. S. Lewis

A Grief Observed
George MacDonald: An Anthology
Mere Christianity
Miracles
The Abolition of Man
The Great Divorce
The Problem of Pain
The Screwtape Letters (with *"Screwtape Proposes a Toast"*)
The Weight of Glory

also available from HarperCollins

The Chronicles of Narnia:
The Magician's Nephew
The Lion, the Witch and the Wardrobe
The Horse and His Boy
Prince Caspian
The Voyage of the Dawn Treader
The Silver Chair
The Last Battle

MIRACLES

A Preliminary Study

C. S. Lewis

HarperOne
An Imprint of HarperCollinsPublishers

To

Cecil and Daphne Harwood

HarperCollins books may be purchased for educational, business, or sales promotional use. For information please e-mail the Special Markets Department at SPsales@harpercollins.com.

HarperCollins website: http://www.harpercollins.com

FIRST HARPERCOLLINS PAPERBACK EDITION PUBLISHED IN 2001

Library of Congress Cataloging-in-Publication Data
Lewis, C. S. (Clive Staples), 1898–1963.
Miracles : a preliminary study / by C. S. Lewis.
p. cm.
Originally published: London : G. Bles, 1947.
ISBN 978-0-06-065301-9
1. Miracles. I. Title.
BT97.2.L49 2000
231.7'3—dc21 00-049863

23 24 25 26 27 LBC 73 72 71 70 69

Among the hills a meteorite
Lies huge; and moss has overgrown,
And wind and rain with touches light
Made soft, the contours of the stone.

Thus easily can Earth digest
A cinder of sidereal fire,
And make her translunary guest
The native of an English shire.

Nor is it strange these wanderers
Find in her lap their fitting place,
For every particle that's hers
Came at the first from outer space.

All that is Earth has once been sky;
Down from the sun of old she came,
Or from some star that travelled by
Too close to his entangling flame.

Hence, if belated drops yet fall
From heaven, on these her plastic power
Still works as once it worked on all
The glad rush of the golden shower.

C.S.L.

Reprinted by permission of *Time and Tide*

CONTENTS

I

THE SCOPE OF THIS BOOK

> Those who wish to succeed must ask the right preliminary questions.
>
> ARISTOTLE, *Metaphysics*, II, (III), I.

In all my life I have met only one person who claims to have seen a ghost. And the interesting thing about the story is that that person disbelieved in the immortal soul before she saw the ghost and still disbelieves after seeing it. She says that what she saw must have been an illusion or a trick of the nerves. And obviously she may be right. Seeing is not believing.

For this reason, the question whether miracles occur can never be answered simply by experience. Every event which might claim to be a miracle is, in the last resort, something presented to our senses, something seen, heard, touched, smelled, or tasted. And our senses are not infallible. If anything extraordinary seems to have happened, we can always say that we have been the victims of an illusion. If we hold a philosophy which excludes the supernatural, this is what we always shall say. What we learn from experience depends on the kind of philosophy we bring to experience. It is therefore useless to appeal to experience before we have settled, as well as we can, the philosophical question.

If immediate experience cannot prove or disprove the miraculous, still less can history do so. Many people think one can decide whether a miracle occurred in the past by examining the evidence 'according to the ordinary rules of historical inquiry'. But the ordinary rules cannot be worked until we have decided whether miracles are possible, and if so, how probable they are. For if they are impossible, then no amount of historical evidence will convince us. If they are possible but immensely improbable, then only mathematically demonstrative evidence will convince us: and since history never provides that degree of evidence for any event, history can never convince us that a miracle occurred. If, on the other hand, miracles are not intrinsically improbable, then the existing evidence will be sufficient to convince us that quite a number of miracles have occurred. The result of our historical enquiries thus depends on the philosophical views which we have been holding before we even began to look at the evidence. This philosophical question must therefore come first.

Here is an example of the sort of thing that happens if we

omit the preliminary philosophical task, and rush on to the historical. In a popular commentary on the Bible you will find a discussion of the date at which the Fourth Gospel was written. The author says it must have been written after the execution of St Peter, because, in the Fourth Gospel, Christ is represented as predicting the execution of St Peter. 'A book', thinks the author, 'cannot be written *before* events which it refers to'. Of course it cannot—unless real predictions ever occur. If they do, then this argument for the date is in ruins. And the author has not discussed at all whether real predictions are possible. He takes it for granted (perhaps unconsciously) that they are not. Perhaps he is right: but if he is, he has not discovered this principle by historical inquiry. He has brought his disbelief in predictions to his historical work, so to speak, ready made. Unless he had done so his historical conclusion about the date of the Fourth Gospel could not have been reached at all. His work is therefore quite useless to a person who wants to know *whether* predictions occur. The author gets to work only after he has already answered that question in the negative, and on grounds which he never communicates to us.

This book is intended as a preliminary to historical inquiry. I am not a trained historian and I shall not examine the historical evidence for the Christian miracles. My effort is to put my readers in a position to do so. It is no use going to the texts until we have some idea about the possibility or probability of the miraculous. Those who assume that miracles cannot happen are merely wasting their time by looking into the texts: we know in advance what results they will find for they have begun by begging the question.

2

THE NATURALIST AND THE SUPERNATURALIST

'Gracious!' exclaimed Mrs Snip, 'and is there a place where people venture to live above ground?' 'I never heard of people living *under* ground," replied Tim, 'before I came to Giant-Land'. 'Came to Giant-Land!' cried Mrs Snip, 'why, isn't everywhere Giant-Land?'

ROLAND QUIZZ, *Giant-Land*, chap xxxii.

I use the word *Miracle* to mean an interference with Nature by supernatural power.[1] Unless there exists, in addition to Nature, something else which we may call the supernatural, there can be no miracles. Some people believe

[1] This definition is not that which would be given by many theologians. I am adopting it not because I think it an improvement upon theirs but precisely because, being crude and 'popular', it enables me most easily to treat those questions which 'the common reader' probably has in mind when he takes up a book on Miracles.

that nothing exists except Nature; I call these people *Naturalists*. Others think that, besides Nature, there exists something else: I call them *Supernaturalists*. Our first question, therefore, is whether the Naturalists or the Supernaturalists are right. And here comes our first difficulty.

Before the Naturalist and the Supernaturalist can begin to discuss their difference of opinion, they must surely have an agreed definition both of Nature and of Supernature. But unfortunately it is almost impossible to get such a definition. Just because the Naturalist thinks that nothing but Nature exists, the word *Nature* means to him merely 'everything' or 'the whole show' or 'whatever there is'. And if that is what we mean by Nature, then of course nothing else exists. The real question between him and the Supernaturalist has evaded us. Some philosophers have defined Nature as 'What we perceive with our five senses'. But this also is unsatisfactory; for we do not perceive our own emotions in that way, and yet they are presumably 'natural' events. In order to avoid this deadlock and to discover what the Naturalist and the Supernaturalist are really differing about, we must approach our problem in a more roundabout way.

I begin by considering the following sentences (1) Are those his natural teeth or a set? (2) The dog in his natural state is covered with fleas. (3) I love to get away from tilled lands and metalled roads and be alone with Nature. (4) Do be natural. Why are you so affected? (5) It may have been wrong to kiss her but it was very natural.

A common thread of meaning in all these usages can easily be discovered. The natural teeth are those which grow in the mouth; we do not have to design them, make them, or fit them. The dog's natural state is the one he will be in if no one takes soap and water and prevents it. The countryside where Nature reigns supreme is the one where soil, weather and vegetation produce their results unhelped and unimpeded by man. Natural behaviour is the behaviour which people would exhibit if they were not at pains to alter it. The natural kiss is the kiss which will be given if moral or prudential considerations do not intervene. In all the examples Nature means what happens 'of itself' or 'of its own accord': what you do not need to labour for; what you will get if you take no measures to stop it. The Greek word for Nature (Physis) is connected with the Greek verb for 'to grow'; Latin *Natura,* with the verb 'to be born'. The Natural is what springs up, or comes forth, or arrives, or goes on, *of its own accord:* the given, what is there already: the spontaneous, the unintended, the unsolicited.

What the Naturalist believes is that the ultimate Fact, the thing you can't go behind, is a vast process in space and time which is *going on of its own accord*. Inside that total system every particular event (such as your sitting reading this book) happens because some other event has happened; in the long run, because the Total Event is happening. Each particular thing (such as this page) is what it is because other things are what they are; and so, eventually, because the whole system is what it is. All the things and events are so completely interlocked that no one of them can claim the slightest independence from 'the whole show'. None of them exists 'on its own' or 'goes on of its own accord' except in the sense that it exhibits, at some particular place and time, that general 'existence on its own' or 'behaviour of its own accord' which belongs to 'Nature' (the great total interlocked event) as a whole. Thus no thoroughgoing Naturalist believes in free will: for free will would mean that human beings have the power of independent action, the power of doing something more or other than what was involved by the total series of events. And any such separate power of originating events is what the Naturalist denies. Spontaneity, originality, action 'on its own', is a privilege reserved for 'the whole show', which he calls *Nature*.

The Supernaturalist agrees with the Naturalist that there must be something which exists in its own right; some basic Fact whose existence it would be nonsensical to try to explain because this Fact is itself the ground or starting-point of all explanations. But he does not identify this Fact with 'the whole show'. He thinks that things fall into two classes. In the first class we find either things or (more probably) One Thing which is basic and original, which exists on its own. In the second we find things which are merely derivative from that One Thing. The one basic Thing has caused all the other things to be. It exists on its own; they exist because it exists. They will cease to exist if it ever ceases to maintain them in existence; they will be altered if it ever alters them.

The difference between the two views might be expressed by saying that Naturalism gives us a democratic, Supernaturalism a monarchical, picture of reality. The Naturalist thinks that the privilege of 'being on its own' resides in the total mass of things, just as in a democracy sovereignty resides in the whole mass of the people. The Supernaturalist thinks that this privilege belongs to some things or (more probably) One Thing and not to others—just as, in a real monarchy, the king has sovereignty and the people have not. And just as, in a democracy, all citizens are equal, so for the Naturalist one thing or event is as good as another, in the sense that they are all equally dependent on the total system of things. Indeed each of them is only the way in which the character of that total system exhibits itself at a particular point in space and time. The Super-naturalist, on the other hand, believes that the one original or self-existent thing is on a different level from, and more important than, all other things.

At this point a suspicion may occur that Supernaturalism first arose from reading into the universe the structure of monarchical societies. But then of course it may with equal reason be suspected that Naturalism has arisen from reading into it the structure of modern democracies. The two suspicions thus cancel out and give us no help in deciding which theory is more likely to be true. They do indeed remind us that Supernaturalism is the characteristic philosophy of a monarchical age and Naturalism of a democratic, in the sense that Supernaturalism, even if false, would have been believed by the great mass of unthinking people four hundred years ago, just as Naturalism, even if false, will be believed by the great mass of unthinking people today.

Everyone will have seen that the One Self-existent Thing—or the small class of self-existent things—in which Supernaturalists believe, is what we call God or the gods. I propose for the rest of this book to treat only that form of Supernaturalism which believes in one God;

partly because polytheism is not likely to be a live issue for most of my readers, and partly because those who believed in many gods very seldom, in fact, regarded their gods as creators of the universe and as self-existent. The gods of Greece were not really supernatural in the strict sense which I am giving to the word. They were products of the total system of things and included within it. This introduces an important distinction.

The difference between Naturalism and Supernaturalism is not exactly the same as the difference between belief in a God and disbelief. Naturalism, without ceasing to be itself, could admit a certain kind of God. The great interlocking event called Nature might be such as to produce at some stage a great cosmic consciousness, an indwelling 'God' arising from the whole process as human mind arises (according to the Naturalists) from human organisms. A Naturalist would not object to that sort of God. The reason is this. Such a God would not stand outside Nature or the total system, would not be existing 'on his own'. It would still be 'the whole show' which was the basic Fact, and such a God would merely be one of the things (even if he were the most interesting) which the basic Fact contained. What Naturalism cannot accept is the idea of a God who stands outside Nature and made it.

We are now in a position to state the difference between the Naturalist and the Supernaturalist despite the fact that they do not mean the same by the word Nature. The Naturalist believes that a great process, of 'becoming', exists 'on its own' in space and time, and that nothing else exists—what we call particular things and events being only the parts into which we analyse the great process or the shapes which that process takes at given moments and given points in space. This single, total reality he calls Nature. The Supernaturalist believes that one Thing exists on its own and has produced the framework of space and time and the procession of systematically connected events which fill them. This framework, and this filling, he calls Nature. It may, or may not, be the only reality which the one Primary Thing has produced. There might be other systems in addition to the one we call Nature.

In that sense there might be several 'Natures'. This conception must be kept quite distinct from what is commonly called 'plurality of worlds'—i.e. different solar systems or different galaxies, 'island universes' existing in widely separated parts of a single space and time. These, however remote, would be parts of the same Nature as our own sun: it and they would be interlocked by being in relations to one another, spatial and temporal relations and casual relations as well. And it is just this reciprocal interlocking within a system which makes it what we call a Nature. Other Natures might not be spatio-temporal at all: or, if any of them were, their space and time would have no spatial or temporal relation to ours. It is just this discontinuity, this failure of interlocking, which would justify us in calling them different Natures. This does not mean that there would be absolutely no relation between them; they would be related by their common derivation from a single Supernatural source. They would, in this respect, be like different novels by a single author; the events in one story have no relation to the events in another *except* that they are invented by the same author. To find the relation between them you must go right back to the author's mind: there is no cutting across from anything Mr Pickwick says in *Pickwick Papers* to anything Mrs Gamp hears in *Martin Chuzzlewit*. Similarly there would be no normal cutting across from an event in one Nature to an event in any other. By a 'normal' relation I mean one which occurs in virtue of the character of the two systems. We have to put in the qualification 'normal' because we do not know in advance that God might not bring two Natures into partial contact at some particular point: that is, He might allow *selected* events in the one to produce results in the other. There would thus be, at certain points, a partial interlocking; but this would not turn the two Natures into one, for the total reciprocity which makes a Nature would still be lacking, and the anomalous interlockings would arise not from what either system was in itself but from the Divine act which was bringing them together. If this occurred each of the two Natures would be 'supernatural' in relation to the other: but the fact of their contact would be supernatural in a more absolute sense—not as being beyond this or that Nature but beyond any and every Nature. It would be one kind of miracle. The other kind would be Divine 'interference' not by the bringing together of two Natures, but simply.

All this is, at present purely speculative. It by no means follows from Supernaturalism that Miracles of any sort do in fact occur. God (the primary thing) may never in fact interfere with the natural system He has created. If He has created more natural systems than one, He may never cause them to impinge on one another.

But that is a question for further consideration. If we decide that Nature is not the only thing there is, then we cannot say in advance whether she is safe from miracles or not. There are things outside her: we do not yet know whether they can get in. The gates may be barred, or they may not. But if Naturalism is true, then we do know in advance that miracles are impossible: nothing can come into Nature from the outside because there is nothing outside to come in, Nature being everything. No doubt, events which we in our ignorance should mistake for miracles might occur: but they would in reality be (just like the commonest events) an inevitable result of the character of the whole system.

Our first choice, therefore, must be between Naturalism and Supernaturalism.

3

THE CARDINAL DIFFICULTY OF NATURALISM

> We cannot have it both ways, and no sneers at the limitations of logic . . . amend the dilemma.
>
> I. A. RICHARDS,
> *Principles of Literary Criticism*, chap. xxv.

If Naturalism is true, every finite thing or event must be (in principle) explicable in terms of the Total System. I say 'explicable *in principle*' because of course we are not going to demand that naturalists, at any given moment, should have found the detailed explanation of every phenomenon. Obviously many things will only be explained when the sciences have made further progress. But if Naturalism is to be accepted we have a right to demand that every single thing should be such that we see, in general, how it could be explained in terms of the Total System. If any one thing exists which is of such a kind that we see in advance the impossibility of ever giving it *that kind* of explanation, then Naturalism would be in ruins. If necessities of thought force us to allow to any one thing any degree of independence from the Total System—if any one thing makes good a claim to be on its own, to be something more than an expression of the character of Nature as a whole—then we have abandoned Naturalism. For by Naturalism we mean the doctrine that only Nature—the whole interlocked system—exists. And if that were true, every thing and event would, if we knew enough, be explicable without remainder (no *heel-taps*) as a necessary product of the system. The whole system being what it is, it ought to be a contradiction in terms if you were not reading this book at the moment; and, conversely, the only cause why you are reading it ought to be that the whole system, at such and such a place and hour, was bound to take that course.

One threat against strict Naturalism has recently been launched on which I myself will base no argument, but which it will be well to notice. The older scientists believed that the smallest particles of matter moved according to strict laws: in other words, that the movements of each particle were 'interlocked' with the total system of Nature. Some modern scientists seem to think—if I understand them—that this is not so. They seem to think that the individual unit of matter (it would be rash to call it any longer a 'particle') moves in an indeterminate or random fashion;

moves, in fact, 'on its own' or 'of its own accord'. The regularity which we observe in the movements of the smallest visible bodies is explained by the fact that each of these contains millions of units and that the law of averages therefore levels out the idiosyncrasies of the individual unit's behaviour. The movement of one unit is incalculable, just as the result of tossing a coin once is incalculable: the majority movement of a billion units can however be predicted, just as, if you tossed a coin a billion times, you could predict a nearly equal number of heads and tails. Now it will be noticed that if this theory is true we have really admitted something other than Nature. If the movements of the individual units are events 'on their own', events which do not interlock with all other events, then these movements are not part of Nature. It would be, indeed, too great a shock to our habits to describe them as *super*-natural. I think we should have to call them *sub*-natural. But all our confidence that Nature has no doors, and no reality outside herself for doors to open on, would have disappeared. There is apparently *something* outside her, the Subnatural; it is indeed from this Subnatural that all events and all 'bodies' are, as it were, fed into her. And clearly if she thus has a back door opening on the Subnatural, it is quite on the cards that she may also have a front door opening on the Supernatural—and events might be fed into her at that door too.

I have mentioned this theory because it puts in a fairly vivid light certain conceptions which we shall have to use later on. But I am not, for my own part, assuming its truth. Those who like myself have had a philosophical rather than a scientific education find it almost impossible to believe that the scientists really mean what they seem to be saying. I cannot help thinking they mean no more than that the movements of individual units are permanently incalculable *to us*, not that they are in themselves random and lawless. And even if they mean the latter, a layman can hardly feel any certainty that some new scientific development may not tomorrow abolish this whole idea of a lawless Subnature. For it is the glory of science to progress. I therefore turn willingly to other ground.

It is clear that everything we know, beyond our own immediate sensations, is inferred from those sensations. I do not mean that we begin as children, by regarding our sensations as 'evidence' and thence arguing consciously to the existence of space, matter, and other people. I mean that if, after we are old enough to understand the question, our confidence in the existence of anything else (say, the solar system or the Spanish Armada) is challenged, our argument in defence of it will have to take the form of inferences from our immediate sensations. Put in its most general form the inference would run, 'Since I am presented with colours, sounds, shapes, pleasures and pains which I cannot perfectly predict or control, and since the more I investigate them the more regular their behaviour appears, therefore there must exist something other than myself and it must be systematic'. Inside this very general inference, all sorts of special trains of inference lead us to more detailed conclusions. We infer Evolution from fossils: we infer the existence of our own brains from what we find inside the skulls of other creatures like ourselves in the dissecting room.

All possible knowledge, then, depends on the validity of reasoning. If the feeling of certainty which we express by words like *must be* and *therefore* and *since* is a real perception of how things outside our own minds really 'must' be, well and good. But if this certainty is merely a feeling *in* our own minds and not a genuine insight into realities beyond them—if it merely represents the way our minds happen to work—then we can have no knowledge. Unless human reasoning is valid no science can be true.

It follows that no account of the universe can be true unless that account leaves it possible for our thinking to be a real insight. A theory which explained everything else in the whole universe but which made it impossible to believe that our thinking was valid, would be utterly out of court. For that theory would itself have been reached by thinking, and if thinking is not valid that theory would, of course, be itself demolished. It would have destroyed its own credentials. It would be an argument which proved that no argument was sound—a proof that there are no such things as proofs—which is nonsense.

Thus a strict materialism refutes itself for the reason given long ago by Professor Haldane: 'If my mental processes are determined wholly by the motions of atoms in my brain, I have no reason to suppose that my beliefs are true . . . and hence I have no reason for supposing my brain to be composed of atoms.' (*Possible Worlds*, p. 209)

But Naturalism, even if it is not purely materialistic, seems to me to involve the same difficulty, though in a somewhat less obvious form. It discredits our processes of reasoning or at least reduces their credit to such a humble level that it can no longer support Naturalism itself.

The easiest way of exhibiting this is to notice the two senses of the word *because*. We can say, 'Grandfather is ill today *because* he ate lobster yesterday.' We can also say, 'Grandfather must be ill today *because* he hasn't got up yet (and we know he is an invariably early riser when he is well).' In the first sentence *because* indicates the relation of Cause and Effect: The eating made him ill. In the second, it indicates the relation of what logicians call Ground and Consequent. The old man's late rising is not the cause of his disorder but the reason why we believe him to be disordered. There is a similar difference between 'He cried out *because* it hurt him' (Cause and Effect) and 'It must have hurt him *because* he cried out' (Ground and Consequent). We are especially familiar with the Ground and Consequent *because* in mathematical reasoning: 'A = C *because*, as we have already proved, they are both equal to B.'

The one indicates a dynamic connection between events or 'states of affairs'; the other, a logical relation between beliefs or assertions.

Now a train of reasoning has no value as a means of finding truth unless each step in it is connected with what went before in the Ground-Consequent relation. If our B does not follow logically from our A, we think in vain. If what we think at the end of our reasoning is to be true, the correct answer to the question, 'Why do you think this?' must begin with the Ground-Consequent *because*.

On the other hand, every event in Nature must be connected with previous events in the Cause and Effect relation. But our acts of thinking are events. Therefore the true answer to 'Why do you think this?' must begin with the Cause-Effect *because*.

Unless our conclusion is the logical consequent from a ground it will be worthless and could be true only by a fluke.

Unless it is the effect of a cause, it cannot occur at all. It looks therefore, as if, in order for a train of thought to have any value, these two systems of connection must apply simultaneously to the same series of mental acts.

But unfortunately the two systems are wholly distinct. To be caused is not to be proved. Wishful thinkings, prejudices, and the delusions of madness, are all caused, but they are ungrounded. Indeed to be caused is so different from being proved that we behave in disputation as if they were mutually exclusive. The mere existence of causes for a belief is popularly treated as raising a presumption that it is groundless, and the most popular way of discrediting a person's opinions is to explain them causally—'You say that *because* (Cause and Effect) you are a capitalist, or a hypochondriac, or a mere man, or only a woman'. The implication is that if causes fully account for a belief, then, since causes work inevitably, the belief would have had to arise whether it had grounds or not. We need not, it is felt, consider grounds for something which can be fully explained without them.

But even if grounds do exist, what exactly have they got to do with the actual occurrence of the belief as a psychological event? If it is an event it must be caused. It must in fact be simply one link in a causal chain which stretches back to the beginning and forward to the end of time. How could such a trifle as lack of logical grounds prevent the belief's occurrence or how could the existence of grounds promote it?

There seems to be only one possible answer. We must say that just as one way in which a mental event causes a subsequent mental event is by Association (when I think of parsnips I think of my first school), so another way in which it can cause it, is simply by being a ground for it. For then being a cause and being a proof would coincide.

But this, as it stands, is clearly untrue. We know by experience that a thought does not necessarily cause all, or even any, of the thoughts which logically stand to it as Consequents to Ground. We should be in a pretty pickle if we could never think 'This is glass' without drawing all the inferences which could be drawn. It is impossible to draw them all; quite often we draw none. We must therefore amend our suggested law. One thought can cause another not by *being,* but by being *seen to be,* a ground for it.

If you distrust the sensory metaphor in *seen,* you may substitute *apprehended* or *grasped* or simply *known.* It makes little difference for all these words recall us to what thinking really is. Acts of thinking are no doubt events; but they are a very special sort of events. They are 'about' something other than themselves and can be true or false. Events in general are not 'about' anything and cannot be true or false. (To say 'these events, or facts are false' means of course that someone's account of them is false). Hence acts of inference can, and must, be considered in two different lights. On the one hand they are subjective events, items in somebody's psychological history. On the other hand, they are insights into, or knowings of, something other than themselves. What from the first point of view is the psychological transition from thought A to thought B, at some particular moment in some particular mind, is, from the thinker's point of view a perception of an implication (if A, then B). When we are adopting the psychological point of view we may use the past tense. 'B *followed* A in my thoughts.' But when we assert the implication we always use the present—'B *follows* from A'. If it ever 'follows from' in the logical sense, it does so always. And we cannot possibly reject the second point of view as a subjective illusion without discrediting all human knowledge. For we can know nothing, beyond our own sensations at the moment unless the act of inference is the real insight that it claims to be.

But it can be this only on certain terms. An act of knowing must be determined, in a sense, solely by what is known; we must know it to be thus solely because it *is* thus. That is what knowing means. You may call this a Cause and Effect *because,* and call 'being known' a mode of causation if you like. But it is a unique mode. The act of knowing has no doubt various conditions, without which it could not occur: attention, and the states of will and health which this presupposes. But its positive character must be determined by the truth it knows. If it were totally explicable from other sources it would cease to be knowledge, just as (to use the sensory parallel) the ringing in my ears ceases to be what we mean by 'hearing' if it can be fully explained from causes other than a noise in the outer world—such as, say, the *tinnitus* produced by a bad cold. If what seems an act of knowledge is partially explicable from other sources, then the knowing (properly so called) in it is just what they leave over, just what demands, for its explanation, the thing known, as real hearing is what is left after you have discounted the *tinnitus*. Any thing which professes to explain our reasoning fully without introducing an act of knowing thus solely determined by what is known, is really a theory that there is no reasoning.

But this, as it seems to me, is what Naturalism is bound to do. It offers what professes to be a full account of our mental behaviour; but this account, on inspection, leaves no room for the acts of knowing or insight on which the whole value of our thinking, as a means to truth, depends.

It is agreed on all hands that reason, and even sentience, and life itself are late comers in Nature. If there is nothing but Nature, therefore, reason must have come into existence by a historical process. And of course, for the Naturalist, this process was not designed to produce a mental behaviour that can find truth. There was no Designer; and indeed, until there were thinkers, there was no truth or falsehood. The type of mental behaviour we now call rational thinking or inference must therefore have been 'evolved' by natural selection, by the gradual weeding out of types less fitted to survive.

Once, then, our thoughts were not rational. That is, all our thoughts once were, as many of our thoughts still are, merely subjective events, not apprehensions of objective truth. Those which had a cause external to ourselves at all were (like our pains) responses to stimuli. Now natural selection could operate only by eliminating responses that were biologically hurtful and multiplying those which tended to survival. But it is not conceivable that any improvement of responses could ever turn them into acts of insight, or even remotely tend to do so. The relation between response and stimulus is utterly different from that between knowledge and the truth known. Our physical vision is a far more useful response to light than that of the cruder organisms which have only a photo-sensitive spot. But neither this improvement nor any possible improvements we can suppose could bring it an inch nearer to being a knowledge of light. It is admittedly something without which we could not have had that knowledge. But the knowledge is achieved by experiments and inferences from them, not by refinement of the response. It is not men with specially good eyes who know about light, but men who have studied the relevant sciences. In the same way our psychological responses to our environment—our curiosities, aversions, delights, expectations—could be indefinitely improved (from the biological point of view) without becoming anything more than responses. Such perfection of the non-rational responses, far from amounting to their conversion into valid inferences, might be conceived as a different method of achieving survival—an alternative to reason. A conditioning which secured that we never felt delight except in the useful nor aversion save from the dangerous, and that the degrees of both were exquisitely proportional to the degree of real utility or danger in the object, might serve us as well as reason or in some circumstances better.

Besides natural selection there is, however, experience—experience originally individual but handed on by tradition and instruction. It might be held that this, in the course of millennia, could conjure the mental behaviour we call reason—in other words, the practice of inference—out of a mental behaviour which was originally not rational.

Repeated experiences of finding fire (or the remains of fire) where he had seen smoke would condition a man to expect fire whenever he saw smoke. This expectation, expressed in the form 'If smoke, then fire' becomes what we call inference. Have all our inferences originated in that way?

But if they did they are all invalid inferences. Such a process will no doubt produce expectation. It will train men to expect fire when they see smoke in just the same way as it trained them to expect that all swans would be white (until they saw a black one) or that water would always boil at 212° (until someone tried a picnic on a mountain). Such expectations are not inferences and need not be true. The assumption that things which have been conjoined in the past will always be conjoined in the future is the guiding principle not of rational but of animal behaviour. Reason comes in precisely when you make the inference 'Since always conjoined, therefore probably connected' and go on to attempt the discovery of the connection. When you have discovered what smoke is you may then be able to replace the mere expectation of fire by a genuine inference. Till this is done reason recognises the expectation as a mere expectation. Where this does not need to be done—that is, where the inference depends on an axiom—we do not appeal to past experience at all. My belief that things which are equal to the same thing are equal to one another is not at all based on the fact that I have never caught them behaving otherwise. I see that it 'must' be so. That some people nowadays call axioms tautologies seems to me irrelevant. It is by means of such 'tautologies' that we advance from knowing less to knowing more. And to call them tautologies is another way of saying that they are completely and certainly known. To see fully that A implies B does (once you have seen it) involve the admission that the assertion of A and the assertion of B are at bottom in the same assertion. The degree to which any true proportion is a tautology depends on the degree of your insight into it. $9 \times 7 = 63$ is a tautology to the perfect arithmetician, but not to the child learning its tables nor to the primitive calculator who reached it, perhaps, by adding seven nines together. If Nature is a totally interlocked system, then every true statement about her (e.g. there was a hot summer in 1959) would be a tautology to an intelligence that could grasp that system in its entirety. 'God is love' may be a tautology to the seraphim; not to men.

'But', it will be said, 'it is incontestable that we do in fact reach truths by inferences'. Certainly. The Naturalist and I both admit this. We could not discuss anything unless we did. The difference I am submitting is that he gives, and I do not, a history of the evolution of reason which is inconsistent with the claims that he and I both have to make for inference as we actually practise it. For his history is, and from the nature of the case can only be, an account, in Cause and Effect terms, of how people came to think the way they do. And this of course leaves in the air the quite different question of how they could possibly be justified in so thinking. This imposes on him the very embarrassing task of trying to show how the evolutionary product which he has described could also be a power of 'seeing' truths.

But the very attempt is absurd. This is best seen if we consider the humblest and almost the most despairing form in which it could be made. The Naturalist might say, 'Well, perhaps we cannot exactly see—not yet—how natural selection would turn sub-rational mental behaviour into inferences that reach truth. But we are certain that this in fact has happened. For natural selection is bound to preserve and increase useful behaviour. And we also find that our habits of inference are in fact useful. And if they are useful they must reach truth'. But notice what we are doing. Inference itself is on trial: that is, the Naturalist has given an account of what we thought to be our inferences which suggests that they are not real insights at all. We, and he, want to be reassured. And the reassurance turns out to be one more inference (if useful, then true)—as if this inference were not, once we accept his evolutionary picture, under the same suspicion as all the rest. If the value of our reasoning is in doubt, you cannot try to establish it by reasoning. If, as I said above, a proof that there are no proofs is nonsensical, so is a proof that there are proofs. Reason is our starting point. There can be no question either of attacking or defending it. If by treating it as a mere phenomenon you put yourself outside it, there is then no way, except by begging the question, of getting inside again.

A still humbler position remains. You may, if you like, give up all claim to truth. You may say simply 'Our way of thinking is useful'—without adding, even under your breath, 'and therefore true'. It enables us to set a bone and build a bridge and make a Sputnik. And that is good enough. The old, high pretensions of reason must be given up. It is a behaviour evolved entirely as an aid to practice. That is why, when we use it simply for practice, we get along pretty well; but when we fly off into speculation and try to get general views of 'reality' we end in the endless, useless, and probably merely verbal, disputes of the philosopher. We will be humbler in future. Goodbye to all that. No more theology, no more ontology, no more metaphysics . . .

But then, equally, no more Naturalism. For of course Naturalism is a prime specimen of that towering speculation, discovered from practice and going far beyond experience, which is now being condemned. Nature is not an object that can be presented either to the senses or the imagination. It can be reached only by the most remote inferences. Or not reached, merely approached. It is the hoped for, the assumed, unification in a single interlocked system of all the things inferred from our scientific experiments. More than that, the Naturalist, not content to assert this, goes on to the sweeping negative assertion. 'There is nothing except this'—an assertion surely, as remote from practice, experience, and any conceivable verification as has ever been made since men began to use their reason speculatively. Yet on the present view, the very first step into such a use was an abuse, the perversion of a faculty merely practical, and the source of all chimeras.

On these terms the Theist's position must be a chimera nearly as outrageous as the Naturalist's. (Nearly, not quite; it abstains from the crowning audacity of a huge negative). But the Theist need not, and does not, grant these terms. He is not committed to the view that reason is a comparatively recent development moulded by a process of selection which can select only the biologically useful. For him, reason—the reason of God—is older than Nature, and from it the orderliness of Nature, which alone enables us to know her, is derived. For him, the human mind in the act of knowing is illuminated by the Divine reason. It is set free, in the measure required, from the huge nexus of non-rational causation; free from this to be determined by the truth known. And the preliminary processes within Nature which led up to this liberation, if there were any, were designed to do so.

To call the act of knowing—the act, not of remembering that something was so in the past, but of 'seeing' that it must be so always and in any possible world—to call this act 'supernatural', is some violence to our ordinary linguistic usage. But of course we do not mean by this that it is spooky, or sensational, or even (in any religious sense) 'spiritual'. We mean only that it 'won't fit in'; that such an act, to be what it claims to be—and if it is not, all our thinking is discredited—cannot be merely the exhibition at a particular place and time of that total, and largely mindless, system of events called 'Nature'. It must break sufficiently free from that universal chain in order to be determined by what it knows.

It is of some importance here to make sure that, if vaguely spatial imagery intrudes (and in many minds it certainly will), it should not be of the wrong kind. We had better not envisage our acts of reason as something 'above' or 'behind' or 'beyond' Nature. Rather 'this side of Nature'—if you must picture spatially, picture them between us and her. It is by inferences that we build up the idea of Nature at all. Reason is given before Nature and on reason our concept of Nature depends. Our acts of inference are prior to our picture of Nature almost as the telephone is prior to the friend's voice we hear by it. When we try to fit these acts into the picture of Nature we fail. The item which we put into that picture and label 'Reason' always turns out to be somehow different from the reason we ourselves are enjoying and exercising while we put it in. The description we have to give of thought as an evolutionary phenomenon always makes a tacit exception in favour of the thinking which we ourselves perform at that moment. For the one can only, like any other particular feat, exhibit, at particular moments in particular consciousnesses, the general and for the most part non-rational working of the whole interlocked system. The other, our present act, claims and must claim, to be an act of insight, a knowledge sufficiently free from non-rational causation to be determined (positively) only by the truth it knows. But the imagined thinking which we put into the picture depends—because our whole idea of Nature depends—on the thinking we are actually doing, not vice versa. This is the prime reality, on which the attribution of reality to anything else rests. If it won't fit into Nature, we can't help it. We will certainly not, on that account, give it up. If we do, we should be giving up Nature too.

4

NATURE AND SUPERNATURE

> Throughout the long tradition of European thought it has been said, not by everyone but by most people, or at any rate by most of those who have proved that they have a right to be heard, that Nature, though it is a thing that really exists, is not a thing that exists in itself or in its own right, but a thing which depends for its existence upon something else.
>
> R. G. COLLINGWOOD,
> *The Idea of Nature*, III iii.

If our argument has been sound, acts of reasoning are not interlocked with the total interlocking system of Nature as all its other items are interlocked with one another. They are connected with it in a different way; as the understanding of a machine is certainly connected with the machine but not in the way the parts of the machine are connected with each other. The knowledge of a thing is not one of the thing's parts. In this sense something beyond Nature operates whenever we reason. I am not maintaining that consciousness as a whole must necessarily be put in the same position. Pleasures, pains, fears, hopes, affections and mental images need not. No absurdity would follow from regarding them as parts of Nature. The distinction we have to make is not one between 'mind' and 'matter', much less between 'soul' and 'body' (hard words, all four of them) but between Reason and Nature: the frontier coming not where the 'outer world' ends and what I should ordinarily call 'myself' begins, but between reason and the whole mass of non-rational events whether physical or psychological.

At that frontier we find a great deal of traffic but it is all one-way traffic. It is a matter of daily experience that rational thoughts induce and enable us to alter the course of Nature—of physical nature when we use mathematics to build bridges, or of psychological nature when we apply arguments to alter our own emotions. We succeed in modifying physical nature more often and more completely than we succeed in modifying psychological nature, but we do at least a little to both. On the other hand, Nature is quite powerless to produce rational thought: not that she never modifies our thinking but that the moment she does so, it ceases (for that very reason) to be rational. For, as we have seen, a train of thought loses all rational credentials as soon as it can be shown to be wholly the result of non-rational causes. When Nature, so to speak, attempts to do things to rational thoughts she only succeeds in killing them. That is the peculiar state of affairs at the frontier. Nature can only raid Reason to kill; but Reason can invade Nature to take prisoners and even to colonise. Every object you see before you at this moment—the walls, ceiling, and furniture, the book, your own washed hands and cut fingernails, bears witness to the colonisation of Nature by Reason: for none of this matter would have been in these states if Nature had had her way. And if you are attending to my argument as closely as I hope, that attention also results from habits which Reason has imposed on the natural ramblings of consciousness. If, on the other hand, a toothache or an anxiety is at this very moment preventing you from attending, then Nature is indeed interfering with your consciousness: but not to produce some new variety of reasoning, only (as far as in her lies) to suspend Reason altogether.

In other words the relation between Reason and Nature is what some people call an Unsymmetrical Relation. Brotherhood is a symmetrical relation because if A is the brother of B, B is the brother of A. Father-and-son is an unsymmetrical relation because if A is the father of B, B is *not* the father of A. The relation between Reason and Nature is of this kind. Reason is not related to Nature as Nature is related to Reason.

I am only too well aware how shocking those who have been brought up to Naturalism will find the picture which begins to show itself. It is, frankly, a picture in which Nature (at any rate on the surface of our own planet) is perforated or pock-marked all over by little orifices at each of which something of a different kind from herself—namely reason—can do things to her. I can only beg you, before you throw the book away, to consider seriously whether your instinctive repugnance to such a conception is really rational, or whether it is only emotional or aesthetic. I know that the hankering for a universe which is all of a piece, and in which everything is the same sort of thing as everything else—a continuity, a seamless web, a democratic universe—is very deep-seated in the modern heart: in mine, no less than in yours. But have we any real assurance that things are like that? Are we mistaking for an intrinsic probability what is really a human desire for tidiness and harmony? Bacon warned us long ago that 'the human understanding is of its own nature prone to suppose the existence of more order and regularity in the world than it finds. And though there be many things which are singular and unmatched, yet it devises for them parallels and conjugates and relatives which do not exist. Hence the fiction that all celestial bodies move in perfect circles' (*Novum Organum*, I, 45). I think Bacon was right. Science itself has already made reality appear less homogeneous than we expected it to be: Newtonian atomism was much more the sort of thing we expected (and desired) than Quantum physics.

If you can, even for the moment, endure the suggested picture of Nature, let us now consider the other factor—the Reasons, or instances of Reason, which attack her. We have seen that rational thought is not part of the system of Nature. Within each man there must be an area (however small) of activity which is outside or independent of her. In relation to Nature, rational thought goes on 'of its own accord' or exists 'on its own'. It does not follow that rational thought exists *absolutely* on its own. It might be independent of Nature by being dependent on something else. For it is not dependence simply but dependence on the non-rational which undermines the credentials of thought. One man's reason has been led to see things by the aid of another man's reason, and is none the worse for that. It is thus still an open question whether each man's reason exists absolutely on its own or whether it is the result of some (rational) cause—in fact, of some other Reason. That other Reason might conceivably be found to depend on a third, and so on; it would not matter how far this process was carried provided you found Reason coming from Reason at each stage. It is only when you are asked to believe in Reason coming from non-reason that you must cry Halt, for, if you don't, all thought is discredited. It is therefore obvious that sooner or later you must admit a Reason which exists absolutely on its own. The problem is whether you or I can be such a self-existent Reason.

This question almost answers itself the moment we remember what existence 'on one's own' means. It means that kind of existence which Naturalists attribute to 'the whole show' and Supernaturalists attribute to God. For instance, what exists on its own must have existed from all eternity; for if anything else could make it begin to exist then it would not exist on its own but because of something else. It must also exist incessantly: that is, it cannot cease to exist and then begin again. For having once ceased to be, it obviously could not recall itself to existence, and if anything else recalled it it would then be a dependent being. Now it is clear that my Reason has grown up gradually since my birth and is interrupted for several hours each night. I therefore cannot be that eternal self-existent Reason which neither slumbers nor sleeps. Yet if any thought is valid, such a Reason must exist and must be the source of my own imperfect and intermittent rationality.

Human minds, then, are not the only supernatural entities that exist. They do not come from nowhere. Each has come into Nature from Supernature: each has its tap-root in an eternal, self-existent, rational Being, whom we call God. Each is an offshoot, or spearhead, or incursion of that Supernatural reality into Nature.

Some people may here raise the following question. If Reason is sometimes present in my mind and sometimes not, then, instead of saying that 'I' am a product of eternal Reason, would it not be wiser to say simply that eternal Reason itself occasionally works through my organism, leaving me a merely natural being? A wire does not become something other than a wire because an electric current has passed through it. But to talk thus is, in my opinion, to forget what reasoning is like. It is not an object which knocks against us, nor even a sensation which we feel. Reasoning doesn't 'happen to' us: we *do* it. Every train of thought is accompanied by what Kant called 'the *I think*'. The traditional doctrine that I am a creature to whom God has given reason but who is distinct from God seems to me much more philosophical than the theory that what appears to be my thinking is only God's thinking through me. On the latter view it is very difficult to explain what happens when I think correctly but reach a false conclusion because I have been misinformed about facts. Why God—who presumably knows the real facts—should be at the pains to think one of His perfectly rational thoughts through a mind in which it is bound to produce error, I do not understand. Nor indeed do I understand why, if all 'my' valid thinking is really God's, He should either Himself mistake it for mine or cause me to mistake it for mine. It seems much more likely that human thought is not God's but God-kindled.

I must hasten, however, to add that this is a book about miracles, not about everything. I am attempting no full doctrine of man:[1] and I am not in the least trying to smuggle in an argument for the 'immortality of the soul'. The earliest Christian documents give a casual and unemphatic assent to the belief that the supernatural part of a man survives the death of the natural organism. But they are very little interested in the matter. What they are intensely interested in is the restoration or 'resurrection' of the whole composite creature by a miraculous divine act: and until we have come to some conclusion about miracles in general we shall certainly not discuss that. At this stage the supernatural element in man concerns us solely as evidence that something beyond Nature exists. The dignity and destiny

[1] See Appendix A.

of man have, at present, nothing to do with the argument. We are interested in man only because his rationality is the little tell-tale rift in Nature which shows that there is something beyond or behind her.

In a pond whose surface was completely covered with scum and floating vegetation, there might be a few water-lilies. And you might of course be interested in them for their beauty. But you might also be interested in them because from their structure you could deduce that they had stalks underneath which went down to roots in the bottom. The Naturalist thinks that the pond (Nature—the great event in space and time) is of an indefinite depth—that there is nothing but water however far you go down. My claim is that some of the things on the surface (i.e. in our experience) show the contrary. These things (rational minds) reveal, on inspection, that they at least are not floating but attached by stalks to the bottom. Therefore the pond has a bottom. It is not pond, pond for ever. Go deep enough and you will come to something that is not pond—to mud and earth and then to rock and finally the whole bulk of Earth and the subterranean fire.

At this point it is tempting to try whether Naturalism cannot still be saved. I pointed out in Chapter II that one could remain a Naturalist and yet believe in a certain kind of God—a cosmic consciousness to which 'the whole show' somehow gave rise: what we might call an *Emergent* God. Would not an Emergent God give us all we need? Is it really necessary to bring in a *super*-natural God, distinct from and outside the whole interlocked system? (Notice, Modern Reader, how your spirits rise—how much more at home you would feel with an emergent, than with a transcendent, God—how much less primitive, repugnant, and naïf the emergent conception seems to you. For by that, as you will see later, there hangs a tale).

But I am afraid it will not do. It is, of course, possible to suppose that when all the atoms of the universe got into a certain relation (which they were bound to get into sooner or later) they would give rise to a universal consciousness. And it might have thoughts. And it might cause those thoughts to pass through our minds. But unfortunately its own thoughts, on this supposition, would be the product of non-rational causes and therefore, by the rule which we use daily, they would have no validity. This cosmic mind would be, just as much as our own minds, the product of mindless Nature. We have not escaped from the difficulty, we have only put it a stage further back. The cosmic mind will help us only if we put it at the beginning, if we suppose it to be, not the product of the total system, but the basic, original, self-existent Fact which exists in its own right. But to admit *that* sort of cosmic mind is to admit a God outside

Nature, a transcendent and supernatural God. This route, which looked like offering an escape, really leads us round again to the place we started from.

There is, then, a God who is not a part of Nature. But nothing has yet been said to show that He must have created her. Might God and Nature be both self-existent and totally independent of each other? If you thought they were you would be a Dualist and would hold a view which I consider manlier and more reasonable than any form of Naturalism. You might be many worse things than a Dualist, but I do not think Dualism is true. There is an enormous difficulty in conceiving two things which simply co-exist and have no other relation. If this difficulty sometimes escapes our notice, that is because we are the victims of picture-thinking. We really imagine them side by side in some kind of space. But of course if they were both in a common space, or a common time, or in any kind of common medium whatever, they would both be parts of a system, in fact of a 'Nature'. Even if we succeed in eliminating such pictures, the mere fact of our trying to think of them together slurs over the real difficulty because, for that moment anyway, our own mind is the common medium. If there can be such a thing as sheer 'otherness', if things can co-exist and no more, it is at any rate a conception which my mind cannot form. And in the present instance it seems specially gratuitous to try to form it, for we already know that God and Nature have come into a certain relation. They have, at the very least, a relation—almost, in one sense, a common frontier—in every human mind.

The relations which arise at that frontier are indeed of a most complicated and intimate sort. That spearhead of the Supernatural which I call my reason links up with all my natural contents—my sensations, emotions, and the like—so completely that I call the mixture by the single word 'me'. Again, there is what I have called the unsymmetrical character of the frontier relations. When the physical state of the brain dominates my thinking, it produces only disorder. But my brain does not become any less a brain when it is dominated by Reason: nor do my emotions and sensations become any the less emotions and sensations. Reason saves and strengthens my whole system, psychological and physical, whereas that whole system, by rebelling against Reason, destroys both Reason and itself. The military metaphor of a spearhead was apparently ill-chosen. The supernatural Reason enters my natural being not like a weapon—more like a beam of light which illuminates or a principle of organisation which unifies and develops. Our whole picture of Nature being 'invaded' (as if by a foreign enemy) was wrong. When we actually examine one of these invasions it looks much more like the arrival of a king among his own subjects or a mahout visiting his own elephant. The elephant may run amuck, Nature may be rebellious. But from observing what happens when Nature obeys it is almost impossible not to conclude that it is her very 'nature' to be a subject. All happens *as if* she had been designed for that very role.

To believe that Nature produced God, or even the human mind, is, as we have seen, absurd. To believe that the two are both independently self-existent is impossible: at least the attempt to do so leaves me unable to say that I am thinking of anything at all. It is true that Dualism has a certain theological attraction; it seems to make the problem of evil easier. But if we cannot, in fact, think Dualism out to the end, this attractive promise can never be kept, and I think there are better solutions of the problem of evil. There remains, then, the belief that God created Nature. This at once supplies a relation between them and gets rid of the difficulty of sheer 'otherness'. This also fits in with the observed frontier situation, in which everything looks as if Nature were not resisting an alien invader but rebelling against a lawful sovereign. This, and perhaps this alone, fits in with the fact that Nature, though not apparently intelligent, is intelligible—that events in the remotest parts of space appear to obey the laws of rational thought. Even the act of creation itself presents none of the intolerable difficulties which seem to meet us on every other hypothesis. There is in our own human minds something that bears a faint resemblance to it. We can imagine: that is, we can cause to exist the mental pictures of material objects, and even human characters, and events. We fall short of creation in two ways. In the first place we can only re-combine elements borrowed from the real universe: no one can imagine a new primary colour or a sixth sense. In the second place, what we imagine exists only for our own consciousness—though we can, by words, induce other people to build for themselves pictures in their own minds which may be roughly similar to it. We should have to attribute to God the power both of producing the basic elements, of inventing not only colours but colour itself, the senses themselves, space, time and matter themselves, and also of imposing what He has invented on created minds. This seems to me no intolerable assumption. It is certainly easier than the idea of God and Nature as wholly unrelated entities, and far easier than the idea of Nature producing valid thought.

I do not maintain that God's creation of Nature can be proved as rigorously as God's existence, but it seems to me overwhelmingly probable, so probable that no one who approached the question with an open mind would very seriously entertain any other hypothesis. In fact one seldom meets people who have grasped the existence of a supernatural God and yet deny that He is the Creator. All the evidence we have points in that direction, and difficulties spring up on every side if we try to believe otherwise. No philosophical theory which I have yet come across is a radical improvement on the words of Genesis, that 'In the beginning God made Heaven and Earth'. I say 'radical' improvement, because the story in Genesis—as St Jerome said long ago—is told in the manner 'of a popular poet', or as we should say, in the form of folk tale. But if you compare it with the creation legends of other peoples—with all these delightful absurdities in which giants to be cut up and floods to be dried up are made to exist *before* creation—the depth and originality of this Hebrew folk tale will soon be apparent. The idea of *creation* in the rigorous sense of the word is there fully grasped.

5

A FURTHER DIFFICULTY IN NATURALISM

> Even as rigorous a determinist as Karl Marx, who at times described the social behaviour of the bourgeoisie in terms which suggested a problem in social physics, could subject it at other times to a withering scorn which only the presupposition of moral responsibility could justify.
>
> R. NIEBUHR, *An Interpretation of Christian Ethics,* chap. iii.

Some people regard logical thinking as the deadest and driest of our activities and may therefore be repelled by the privileged position I gave it in the last chapter. But logical thinking—Reasoning—had to be the pivot of the argument because, of all the claims which the human mind puts forward, the claim of Reasoning to be valid is the only one which the Naturalist cannot deny without (philosophically speaking) cutting his own throat. You cannot, as we saw, prove that there are no proofs. But you can if you wish regard all human ideals as illusions and all human loves as biological by-products. That is, you can do so without running into flat self-contradiction and nonsense. Whether you can do so without extreme unplausibility—without accepting a picture of things which no one really believes—is another matter.

Besides reasoning about matters of fact, men also make moral judgements—'I ought to do this'—'I ought not to do that'—'This is good'—'That is evil.' Two views have been held about moral judgements. Some people think that when we make them we are not using our Reason, but are employing some different power. Other people think that we make them by our Reason. I myself hold this second view. That is, I believe that the primary moral principles on which all others depend are rationally perceived. We 'just see' that there is no reason why my neighbour's happiness should be sacrificed to my own, as we 'just see' that things which are equal to the same thing are equal to one another. If we cannot prove either axiom, that is not because they are irrational but because they are self-evident and all proofs depend on them. Their intrinsic reasonableness shines by its own light. It is because all morality is based on such self-evident principles that we say to a man, when we would recall him to right conduct, 'Be reasonable.'

But this is by the way. For our present purpose it does not matter which of these two views you adopt. The important point is to notice that moral judgements raise the same sort of difficulty for Naturalism as any other thoughts. We always assume in discussions about morality, as in all other discussions, that the other man's views are worthless if they can be fully accounted for by some non-moral and non-rational cause. When two men differ about good and evil we soon hear this principle being brought into play. 'He believes in the sanctity of property because he's a millionaire'—'He believes in Pacifism because he's a coward'—'He approves of corporal punishment because he's a sadist.' Such taunts may often be untrue: but the mere fact that they are made by the one side, and hotly rebutted by the other, shows clearly what principle is being used. Neither side doubts that if they were true they would be decisive. No one (in real life) pays attention to any moral judgement which can be shown to spring from non-moral and non-rational causes. The Freudian and the Marxist attack traditional morality precisely on this ground—and with wide success. All men accept the principle.

But, of course, what discredits particular moral judgements must equally discredit moral judgement as a whole. If the fact that men have such ideas as *ought* and *ought not* at all can be fully explained by irrational and non-moral causes, then those ideas are an illusion. The Naturalist is ready to explain how the illusion arose. Chemical conditions produce life. Life, under the influence of natural selection, produces consciousness. Conscious organisms which behave in one way live longer than those which behave in another. Living longer, they are more likely to have offspring. Inheritance, and sometimes teaching as well, pass on their mode of behaviour to their young. Thus in every species a pattern of behaviour is built up. In the human species conscious teaching plays a larger part in building it up, and the tribe further strengthens it by killing individuals who don't conform. They also invent gods who are said to punish departures from it. Thus, in time, there comes to exist a strong human impulse to conform. But since this impulse is often at variance with the other impulses, a mental conflict arises, and the man expresses it by saying 'I want to do A but I ought to do B.'

This account may (or may not) explain why men do in fact make moral judgements. It does not explain how they could be right in making them. It excludes, indeed, the very possibility of their being right. For when men say 'I ought' they certainly think they are saying something, and something true, about the nature of the proposed action, and not merely about their own feelings. But if Naturalism is true, 'I ought' is the same sort of statement as 'I itch' or 'I'm going to be sick.' In real life when a man says 'I ought' we may reply, 'Yes. You're right. That *is* what you ought to do,' or else, 'No. I think you're mistaken.' But in a world of Naturalists (if Naturalists really remembered their philosophy out of school) the only sensible reply would be, 'Oh, are you?' All moral judgements would be statements about the speaker's feelings, mistaken by him for statements about something else (the real moral quality of actions) which does not exist.

Such a doctrine, I have admitted, is not flatly self-contradictory. The Naturalist can, if he chooses, brazen it out. He can say, 'Yes. I quite agree that there is no such thing as wrong and right. I admit that no moral judgement can be "true" or "correct" and, consequently, that no one system of morality can be better or worse than another. All ideas of good and evil are hallucinations—shadows cast on the outer world by the impulses which we have been conditioned to feel.' Indeed many Naturalists are delighted to say this.

But then they must stick to it; and fortunately (though inconsistently) most real Naturalists do not. A moment after they have admitted that good and evil are illusions, you will find them exhorting us to work for posterity, to educate, revolutionise, liquidate, live and die for the good of the human race. A Naturalist like Mr H. G. Wells spent a long life doing so with passionate eloquence and zeal. But surely this is very odd? Just as all the books about spiral nebulae, atoms and cave men would really have led you to suppose that the Naturalists claimed to be able to know something, so all the books in which Naturalists tell us what we ought to do would really make you believe that they thought some ideas of good (their own, for example) to be somehow preferable to others. For they write with indignation like men proclaiming what is good in itself and denouncing what is evil in itself, and not at all like men recording that they personally like mild beer but some people prefer bitter. Yet if the 'oughts' of Mr Wells and, say, Franco are both equally the impulses which Nature has conditioned each to have and both tell us nothing about any objective right or wrong, whence is all the fervour? Do they remember while they are writing thus that when they tell us we 'ought to make a better world' the words 'ought' and 'better' must, on their own showing, refer to an irrationally conditioned impulse which cannot be true or false any more than a vomit or a yawn?

My idea is that sometimes they do forget. That is their glory. Holding a philosophy which excludes humanity, they yet remain human. At the sight of injustice they throw all their Naturalism to the winds and speak like men and like men of genius. They know far better than they think they know. But at other times, I suspect they are trusting in a supposed way of escape from their difficulty.

It works—or *seems* to work—like this. They say to themselves, 'Ah, yes. Morality'—or 'bourgeois morality' or 'conventional morality' or 'traditional morality' or some such addition—'Morality *is* an illusion. But we have found out what modes of behaviour will in fact preserve the human race alive. That is the behaviour we are pressing you to adopt. Pray don't mistake us for moralists. We are under an entirely new management'... just as if this would help. It would help only if we grant, firstly, that life is better than death and, secondly, that we ought to care for the lives of our descendants as much as, or more than, for our own. And both these are moral judgements which have, like all others, been explained away by Naturalism. Of course, having been conditioned by Nature in a certain way, we do feel thus about life and about posterity. But the Naturalists have cured us of mistaking these feelings for insights into what we once called 'real value'. Now that I know that my impulse to serve posterity is just the same kind of thing as my fondness for cheese—now that its transcendental pretensions have been exposed for a sham—do you think I shall pay much attention to it? When it happens to be strong (and it has grown considerably weaker since you explained to me its real nature) I suppose I shall obey it. When it is weak, I shall put my money into cheese. There can be no reason for trying to whip up and encourage the one impulse rather than the other. Not now that I know what they both are. The Naturalists must not destroy all my reverence for conscience on Monday and expect to find me still venerating it on Tuesday.

There is no escape along those lines. If we are to continue to make moral judgements (and whatever we say we shall in fact continue) then we must believe that the conscience of man is not a product of Nature. It can be valid only if it is an offshoot of some absolute moral wisdom, a moral wisdom which exists absolutely 'on its own' and is not a product of non-moral, non-rational Nature. As the argument of the last chapter led us to acknowledge a supernatural source for rational thought, so the argument of this leads us to acknowledge a supernatural source for our ideas of good and evil. In other words, we now know something more about God. If you hold that moral judgement is a different thing from Reasoning you will express this new knowledge by saying, 'We now know that God has at least one other attribute than rationality.' If, like me, you hold that moral judgement is a kind of Reasoning, then you will say, 'We now know more about the Divine Reason.'

And with this we are almost ready to begin our main argument. But before doing so it will be well to pause for the consideration of some misgivings or misunderstandings which may have already arisen.

6

ANSWERS TO MISGIVINGS

> For as bats' eyes are to daylight so is our intellectual eye to those truths which are, in their own nature, the most obvious of all.
>
> ARISTOTLE, *Metaphysics,* I (Brevior) i.

It must be clearly understood that the argument so far leads to no conception of 'souls' or 'spirits' (words I have avoided) floating about in the realm of Nature with no relation to their environment. Hence we do not deny—indeed we must welcome—certain considerations which are often regarded as proofs of Naturalism. We can admit, and even insist, that Rational Thinking can be shown to be conditioned in its exercise by a natural object (the brain). It is temporarily impaired by alcohol or a blow on the head. It wanes as the brain decays and vanishes when the brain ceases to function. In the same way the moral outlook of a community can be shown to be closely connected with its history, geographical environment, economic structure, and so forth. The moral ideas of the individual are equally related to his general situation: it is no accident that parents and schoolmasters so often tell us that they can stand any vice rather than lying, the lie being the only defensive weapon of the child. All this, far from presenting us with a difficulty, is exactly what we should expect.

The rational and moral element in each human mind is a point of force from the Supernatural working its way into Nature, exploiting at each point those conditions which Nature offers, repulsed where the conditions are hopeless and impeded when they are unfavourable. A man's Rational thinking is *just so much* of his share in eternal Reason as the state of his brain allows to become operative: it represents, so to speak, the bargain struck or the frontier fixed between Reason and Nature at that particular point. A nation's moral outlook is just so much of its share in eternal Moral Wisdom as its history, economics etc. lets through. In the same way the voice of the Announcer is just so much of a human voice as the receiving set lets through. Of course it varies with the state of the receiving set, and deteriorates as the set wears out and vanishes altogether if I throw a brick at it. It is conditioned by the apparatus but not originated by it. If it were—if we knew that there was no human being at the microphone—we should not attend to the news. The various and complex conditions under which Reason and

Morality appear are the twists and turns of the frontier between Nature and Supernature. That is why, if you wish, you can always ignore Supernature and treat the phenomena purely from the Natural side; just as a man studying on a map the boundaries of Cornwall and Devonshire can always say, 'What you call a bulge in Devonshire is really a dent in Cornwall.' And in a sense you can't refute him. What we call a bulge in Devonshire always *is* a dent in Cornwall. What we call rational thought in a man always involves a state of the brain, in the long run a relation of atoms. But Devonshire is none the less something more than 'where Cornwall ends', and Reason is something more than cerebral biochemistry.

I now turn to another possible misgiving. To some people the great trouble about any argument for the Supernatural is simply the fact that argument should be needed at all. If so stupendous a thing exists, ought it not to be obvious as the sun in the sky? Is it not intolerable, and indeed incredible, that knowledge of the most basic of all Facts should be accessible only by wire-drawn reasonings for which the vast majority of men have neither leisure nor capacity? I have great sympathy with this point of view. But we must notice two things.

When you are looking at a garden from a room upstairs it is obvious (once you think about it) that you are looking through a window. But if it is the garden that interests you, you may look at it for a long time without thinking of the window. When you are reading a book it is obvious (once you attend to it) that you are using your eyes: but unless your eyes begin to hurt you, or the book is a text book on optics, you may read all evening without once thinking of eyes. When we talk we are obviously using language and grammar: and when we try to talk a foreign language we may be painfully aware of the fact. But when we are talking English we don't notice it. When you shout from the top of the stairs, 'I'm coming in half a moment,' you are not usually conscious that you have made the singular *am* agree with the singular *I*. There is indeed a story told about a Redskin who, having learned several other languages, was asked to write a grammar of the language used by his own tribe. He replied, after some thought, that it had no grammar. The grammar he had used all his life had escaped his notice all his life. He knew it (in one sense) so well that (in another sense) he did not know it existed.

All these instances show that the fact which is in one respect the most obvious and primary fact, and through which alone you have access to all the other facts, may be precisely the one that is most easily forgotten—forgotten not because it is so remote or abstruse but because it is so near and so obvious. And that is exactly how the Supernatural has been forgotten. The Naturalists have been engaged in thinking about Nature. They have not attended to the fact that they were *thinking*. The moment one attends to this it is obvious that one's own thinking cannot be merely a natural event, and that therefore something other than Nature exists. The Supernatural is not remote and abstruse: it is a matter of daily and hourly experience, as intimate as breathing. Denial of it depends on a certain absent-mindedness. But this absent-mindedness is in no way surprising. You do not need—indeed you do not wish—to be always thinking about windows when you are looking at gardens or always thinking about eyes when you are reading. In the same way the proper procedure for all limited and particular inquiries is to ignore the fact of your own thinking, and concentrate on the object. It is only when you stand back from particular inquiries and try to form a complete philosophy that you must take it into account. For a complete philosophy must get in *all* the facts. In it you turn away from specialised or truncated thought to total thought: and one of the facts total thought must think about is Thinking itself. There is thus a tendency in the study of Nature to make us forget the most obvious fact of all. And since the sixteenth century, when Science was born, the minds of men have been increasingly turned outward, to know Nature and to master her. They have been increasingly engaged on those specialised inquiries for which truncated thought is the correct method. It is therefore not in the least astonishing that they should have forgotten the evidence for the Supernatural. The deeply ingrained habit of truncated thought—what we call the 'scientific' habit of mind—was indeed certain to lead to Naturalism, unless this tendency were continually corrected from some other source. But no other source was at hand, for during the same period men of science were coming to be metaphysically and theologically uneducated.

That brings me to the second consideration. The state of affairs in which ordinary people can discover the Supernatural only by abstruse reasoning is recent and, by historical standards, abnormal. All over the world, until quite modern times, the direct insight of the mystics and the reasonings of the philosophers percolated to the mass of the people by authority and tradition; they could be received by those who were no great reasoners themselves in the concrete form of myth and ritual and the whole pattern of life. In the conditions produced by a century or so of Naturalism, plain men are being forced to bear burdens which plain men were never expected to bear before. We must get the truth for ourselves or go without it. There may be two explanations for this. It might be that humanity, in rebelling against tradition and authority, has made a ghastly mistake; a mistake which will not be the less fatal because the corruptions of those in authority rendered it very excusable. On the other hand, it may be that the Power which rules our species is at this moment carrying out a daring experiment. Could it be intended that the whole mass of the people should now move forward and occupy for themselves those heights which were once reserved only for the sages? Is the distinction between wise and simple to disappear because all are now expected to become wise? If so, our present blunderings would be but growing pains. But let us make no mistake about our necessities. If we are content to go back and become humble plain men obeying a tradition, well. If we are ready to climb and struggle on till we become sages ourselves, better still. But the man who will neither obey wisdom in others nor adventure for her/himself is fatal. A society where the simple many obey the few seers can live: a society where all were seers could live even more fully. But a society where the mass is still simple and the seers are no longer attended to can achieve only superficiality, baseness, ugliness, and in the end extinction. On or back we must go; to stay here is death.

One other point that may have raised doubt or difficulty should here be dealt with. I have advanced reasons for believing that a supernatural element is present in every rational man. The presence of human rationality in the world is therefore a Miracle by the definition given in Chapter II. On realising this the reader may excusably say, 'Oh, if *that's* all he means by a Miracle . . .' and fling the book away. But I ask him to have patience. Human Reason and Morality have been mentioned not as instances of Miracle (at least, not of the kind of Miracle you wanted to hear about) but as proofs of the Supernatural: not in order to show that Nature ever is invaded but that there is a possible invader. Whether you choose to call the regular and familiar invasion by human Reason a Miracle or not is largely a matter of words. Its regularity—the fact that it regularly enters by the same door, human sexual intercourse—may incline you not to do so. It looks as if it were (so to speak) the very nature of Nature to suffer *this* invasion. But then we might later find that it was the very nature of Nature to suffer Miracles in general. Fortunately the course of our argument will allow us to leave this question of terminology on one side. We are going to be concerned with other invasions of Nature—with what everyone would call Miracles. Our question could, if you liked, be put in the form, 'Does Supernature ever produce particular results in space and time *except* through the instrumentality of human brains acting on human nerves and muscles?'

I have said '*particular* results' because, on our view, Nature as a whole is herself one huge result of the

Supernatural: God created her. God pierces her wherever there is a human mind. God presumably maintains her in existence. The question is whether He ever does anything else to her. Does He, besides all this, ever introduce into her events of which it would not be true to say, 'This is simply the working out of the general character which He gave to Nature as a whole in creating her'? Such events are what are popularly called Miracles: and it will be in this sense only that the word Miracle will be used for the rest of the book.

7

A CHAPTER OF RED HERRINGS

> Thence came forth *Maul*, a giant. This *Maul* did use to spoil young Pilgrims with sophistry.
>
> BUNYAN

From the admission that God exists and is the author of Nature, it by no means follows that miracles must, or even can, occur. God Himself might be a being of such a kind that it was contrary to His character to work miracles. Or again, He might have made Nature the sort of thing that cannot be added to, subtracted from, or modified. The case against Miracles accordingly relies on two different grounds. You either think that the character of God excludes them or that the character of Nature excludes them. We will begin with the second which is the more popular ground. In this chapter I shall consider forms of it which are, in my opinion, very superficial—which might even be called misunderstandings or Red Herrings.

The first Red Herring is this. Any day you may hear a man (and not necessarily a disbeliever in God) say of some alleged miracle, 'No. Of course I don't believe that. We know it is contrary to the laws of Nature. People could believe it in olden times because they didn't know the laws of Nature. We know now that it is a scientific impossibility'.

By the 'laws of Nature' such a man means, I think, the observed course of Nature. If he means anything more than that he is not the plain man I take him for but a philosophic Naturalist and will be dealt with in the next chapter. The man I have in view believes that mere experience (and specially those artificially contrived experiences which we call Experiments) can tell us what regularly happens in Nature. And he thinks that what we have discovered excludes the possibility of Miracle. This is a confusion of mind.

Granted that miracles *can* occur, it is, of course, for experience to say whether one has done so on any given occasion. But mere experience, even if prolonged for a million years, cannot tell us whether the thing is possible. Experiment finds out what regularly happens in Nature: the norm or rule to which she works. Those who believe in miracles are not denying that there is such a norm or rule: they are only saying that it can be suspended. A miracle is by definition an exception. How can the discovery of the rule tell you whether, granted a sufficient cause, the rule can be suspended? If we said that the rule was A, then experience might refute us by discovering that it was B. If we said that there was no rule, then experience might refute us by observing that there is. But we are saying neither of these things. We agree that there is a rule and that the rule is B. What has that got to do with the question whether the rule can be suspended? You reply, 'But experience shows that it never has'. We reply, 'Even if that were so, this would not prove that it never can. But does experience show that it never has? The world is full of stories of people who say they have experienced miracles. Perhaps the stories are false: perhaps they are true. But before you can decide on that historical question, you must first (as was pointed out in Chapter 1) discover whether the thing is possible, and if possible, how probable'.

The idea that the progress of science has somehow altered this question is closely bound up with the idea that people 'in olden times' believe in them 'because they didn't know the laws of Nature'. Thus you will hear people say, 'The early Christians believed that Christ was the son of a virgin, but we know that this is a scientific impossibility'. Such people seem to have an idea that belief in miracles arose at a period when men were so ignorant of the course of nature that they did not perceive a miracle to be contrary to it. A moment's thought shows this to be nonsense: and the story of the Virgin Birth is a particularly striking example. When St Joseph discovered that his fiancée was going to have a baby, he not unnaturally decided to repudiate her. Why? Because he knew just as well as any modern gynaecologist that in the ordinary course of nature women do not have babies unless they have lain with men. No doubt the modern gynaecologist knows several things about birth and begetting which St Joseph did not know. But those things do not concern the main point—that a virgin birth is contrary to the course of nature. And St Joseph obviously knew *that*. In any sense in which it is true to say now, 'The thing is scientifically impossible', he would have said the same: the thing always was, and was always known to be, impossible *unless* the regular processes of nature were, in this particular case, being over-ruled or supplemented by something from beyond nature. When St Joseph finally accepted the view that his fiancée's pregnancy was due not to unchastity but to a miracle, he accepted the miracle as something contrary to the known order of nature. All records of miracles teach the same thing. In such stories the miracles excite fear and wonder (that is what the very word *miracle* implies) among the spectators, and are taken as evidence of supernatural power. If they were not known to be contrary to the laws of nature how could they suggest the presence of the supernatural? How could they be surprising unless they were seen to be exceptions to the rules? And how can anything be seen to be an exception till the rules are known? If there ever were men who did not know the laws of nature *at all,* they would have no idea of a miracle and feel no particular interest in one if it were performed before them. Nothing can seem extraordinary until you have discovered what is ordinary. Belief in miracles, far from depending on an ignorance of the laws of nature, is only possible in so far as those laws are known. We have already seen that if you begin by ruling out the supernatural you will perceive no miracles. We must now add that you will equally perceive no miracles until you believe that nature works according to regular laws. If you have not yet noticed that the sun always rises in the East you will see nothing miraculous about his rising one morning in the West.

If the miracles were offered us as events that normally occurred, then the progress of science, whose business is to tell us what normally occurs, would render belief in them gradually harder and finally impossible. The progress of science has in just this way (and greatly to our benefit) made all sorts of things incredible which our ancestors believed; man-eating ants and gryphons in Scythia, men with one single gigantic foot, magnetic islands that draw all ships towards them, mermaids and fire-breathing dragons.

But those things were never put forward as supernatural interruptions of the course of nature. They were put forward as items within her ordinary course—in fact as 'science'. Later and better science has therefore rightly removed them. Miracles are in a wholly different position. If there were fire-breathing dragons our big-game hunters would find them: but no one ever pretended that the Virgin Birth or Christ's walking on the water could be reckoned on to recur. When a thing professes from the very outset to be a unique invasion of Nature by something from outside, increasing knowledge of Nature can never make it either more or less credible than it was at the beginning. In this sense it is mere confusion of thought to suppose that advancing science has made it harder for us to accept miracles. We always knew they were contrary to the natural course of events; we know still that if there is something beyond Nature, they are possible. Those are the bare bones of the question; time and progress and science and civilisation have not altered them in the least. The grounds for belief and disbelief are the same today as they were two thousand—or ten thousand—years ago. If St Joseph had lacked faith to trust God or humility to perceive the holiness of his spouse, he could have disbelieved in the miraculous origin of her Son as easily as any modern man; and any modern man who believes in God can accept the miracle as easily as St Joseph did. You and I may not agree, even by the end of this book, as to whether miracles happen or not. But at least let us not talk nonsense. Let us not allow vague rhetoric about the march of science to fool us into supposing that the most complicated account of birth, in terms of genes and spermatozoa, leaves us any more convinced than we were before that *nature* does not send babies to young women who 'know not a man'.

The second Red Herring is this. Many people say, 'They could believe in miracles in olden times because they had a false conception of the universe. They thought the Earth was the largest thing in it and Man the most important creature. It therefore seemed reasonable to suppose that the Creator was specially interested in Man and might even interrupt the course of Nature for his benefit. But now that we know the real immensity of the universe—now that we perceive our own planet and even the whole Solar System to be only a speck—it becomes ludicrous to believe in them any longer. We have discovered our significance and can no longer suppose that God is so drastically concerned in our petty affairs'.

Whatever its value may be as an argument, it may be stated at once that this view is quite wrong about facts. The immensity of the universe is not a recent discovery. More than seventeen hundred years ago Ptolemy taught that in relation to the distance of the fixed stars the whole Earth must be regarded as a point with no magnitude. His astronomical system was universally accepted in the Dark and Middle Ages. The insignificance of Earth was as much a commonplace to Boethius, King Alfred, Dante, and Chaucer as it is to Mr H. G. Wells or Professor Haldane. Statements to the contrary in modern books are due to ignorance.

The real question is quite different from what we commonly suppose. The real question is why the spatial insignificance of Earth, after being asserted by Christian philosophers, sung by Christian poets, and commented on by Christian moralists for some fifteen centuries, without the slightest suspicion that it conflicted with their theology, should suddenly in quite modern times have been set up as a stock argument against Christianity and enjoyed, in that capacity, a brilliant career. I will offer a guess at the answer to this question presently. For the moment, let us consider the strength of this stock argument.

When the doctor at a post-mortem looks at the dead man's organs and diagnoses poison he has a clear idea of the different state in which the organs would have been if the man had died a natural death. If from the vastness of the universe and the smallness of Earth we diagnose that Christianity is false we ought to have a clear idea of the sort of universe we should have expected if it were true. But have we? Whatever space may really be, it is certain that our perceptions make it appear three dimensional; and to a three-dimensional space no boundaries are conceivable. By the very forms of our perceptions therefore we must feel as if we lived somewhere in infinite space: and whatever size the Earth happens to be, it must of course be very small in comparison with infinity. And this infinite space must either be empty or contain bodies. If it were empty, if it contained nothing but our own Sun, then that vast vacancy would certainly be used as an argument against the very existence of God. Why, it would be asked, should He create one speck and leave all the rest of space to nonentity? If, on the other hand, we find (as we actually do) countless bodies floating in space, they must be either habitable or uninhabitable. Now the odd thing is that *both* alternatives are equally used as objections to Christianity. If the universe is teeming with life other than ours, then this, we are told, makes it quite ridiculous to believe that God should be so concerned with the human race as to 'come down from Heaven' and be made man for its redemption. If, on the other hand, our planet is really unique in harbouring organic life, then this is thought to prove that life is only an accidental by-product in the universe and so again to disprove our religion. We treat God as the policeman in the story treated the suspect; whatever he does 'will be used in evidence against Him'. This kind of objection to the Christian faith is not really based on the observed nature of the actual universe at all. You can make it without waiting to find out what the universe is like, for it will fit any kind of universe we choose to imagine. The doctor here can diagnose poison without looking at the corpse for he has a theory of poison which he will maintain *whatever* the state of the organs turns out to be.

The reason why we cannot even imagine a universe so built as to exclude these objections is, perhaps, as follows. Man is a finite creature who has sense enough to know that he is finite: therefore, on any conceivable view, he finds himself dwarfed by reality as a whole. He is also a derivative being: the cause of his existence lies not in himself but (immediately) in his parents and (ultimately) *either* in the character of Nature as a whole *or* (if there is a God) in God. But there must be something, whether it be God or the totality of Nature, which exists in its own right or goes on 'of its own accord'; not as the product of causes beyond itself, but simply because it does. In the face of that something, whichever it turns out to be, man must feel his own derived existence to be unimportant, irrelevant, almost accidental. There is no question of religious people fancying that all exists for man and scientific people discovering that it does not. Whether the ultimate and inexplicable being—that which simply *is*—turns out to be God or 'the whole show', of course it does not exist for us. On either view we are faced with something which existed before the human race appeared and will exist after the Earth has become uninhabitable; which is utterly independent of us though we are totally dependent on it; and which, through vast ranges of its being, has no relevance to our own hopes and fears. For no man was, I suppose, ever so mad as to think that man, or all creation, *filled* the Divine Mind; if we are a small thing to space and time, space and time are a much smaller thing to God. It is a profound mistake to imagine that Christianity ever intended to dissipate the bewilderment and even the terror, the sense of our own nothingness, which come upon us when we think about the nature of things. It comes to intensify them. Without such sensations there is no religion. Many a man, brought up in the glib profession of some shallow form of Christianity, who comes through reading Astronomy to realise for the first time how majestically indifferent most reality is to man, and who perhaps abandons his religion on that account, may at that moment be having his first genuinely religious experience.

Christianity does not involve the belief that all things were made for man. It does involve the belief that God loves man and for his sake became man and died. I have not yet succeeded in seeing how what we know (and have known since the days of Ptolemy) about the size of the universe affects the credibility of this doctrine one way or the other.

The sceptic asks how we can believe that God so 'came down' to this one tiny planet. The question would be embarrassing if we knew (1) that there are rational creatures on any of the other bodies that float in space; (2) that they have, like us, fallen and need redemption; (3) that their redemption must be in the same mode as ours; (4) that redemption in this mode has been withheld from them. But we know none of them. The universe may be full of happy lives that never needed redemption. It may be full of lives that have been redeemed in modes suitable to their condition, of which we can form no conception. It may be full of lives that have been redeemed in the very same mode as our own. It may be full of things quite other than life in which God is interested though we are not.

If it is maintained that anything so small as the Earth must, in any event, be too unimportant to merit the love of the Creator, we reply that no Christian ever supposed we did merit it. Christ did not die for men because they were intrinsically worth dying for, but because He is intrinsically love, and therefore loves infinitely. And what, after all, does the *size* of a world or a creature tell us about its 'importance' or value?

There is no doubt that we all *feel* the incongruity of supposing, say, that the planet Earth might be more important than the Great Nebula in Andromeda. On the other hand, we are all equally certain that only a lunatic would think a man six-feet high necessarily more important than a man five-feet high, or a horse necessarily more important than a man, or a man's legs than his brain. In other words this supposed ratio of size to importance feels plausible only when one of the sizes involved is very great. And that betrays the true basis of this type of thought. When a relation is perceived by Reason, it is perceived to hold good universally. If our Reason told us that size was proportional to importance, then small differences in size would be accompanied by small differences in importance just as surely as great differences in size were accompanied by great differences in importance. Your six-foot man would have to be slightly more valuable than the man of five feet, and your leg slightly more important than your brain—which everyone knows to be nonsense. The conclusion is inevitable: the importance we attach to great differences of size is an affair not of reason but of emotion—of that peculiar emotion which superiorities in size begin to produce in us only after a certain point of absolute size has been reached.

We are inveterate poets. When a quantity is very great we cease to regard it as a mere quantity. Our imaginations awake. Instead of mere quantity, we now have a quality—the Sublime. But for this, the merely arithmetical greatness of the Galaxy would be no more impressive than the figures in an account book. To a mind which did not share our emotions and lacked our imaginative energies, the argument against Christianity from the size of the universe would be simply unintelligible. It is therefore from ourselves that the material universe derives its power to overawe us. Men of sensibility look up on the night sky with awe: brutal and stupid men do not. When the silence of the eternal spaces terrified Pascal, it was Pascal's own greatness that enabled them to do so; to be frightened by the bigness of the nebulae is, almost literally, to be frightened at our own shadow. For light years and geological periods are mere arithmetic until the shadow of man, the poet, the maker of myths, falls upon them. As a Christian I do not say we are wrong to tremble at that shadow, for I believe it to be the shadow of an image of God. But if the vastness of Nature ever threatens to overcrow our spirits, we must remember that it is only Nature spiritualised by human imagination which does so.

This suggests a possible answer to the question raised a few pages ago—why the size of the universe, known for centuries, should first in modern times become an argument against Christianity. Has it perhaps done so because in modern times the imagination has become more sensitive to bigness? From this point of view the argument from size might almost be regarded as a by-product of the Romantic Movement in poetry. In addition to the absolute increase of imaginative vitality on this topic, there has pretty certainly been a decline on others. Any reader of old poetry can see that brightness appealed to ancient and medieval man more than bigness, and more than it does to us. Medieval thinkers believed that the stars must be somehow superior to the Earth because they looked bright and it did not. Moderns think that the Galaxy ought to be more important than the Earth because it is bigger. Both states of mind can produce good poetry. Both can supply mental pictures which rouse very respectable emotions—emotions of awe, humility, or exhilaration. But taken as serious philosophical argument both are ridiculous. The atheist's argument from size is, in fact, an instance of just that picture-thinking to which, as we shall see in a later chapter, the Christian is *not* committed. It is the particular mode in which picture-thinking appears in the twentieth century: for what we fondly call 'primitive' errors do not pass away. They merely change their form.

8

MIRACLES AND THE LAWS OF NATURE

It's a very odd thing –
As odd as can be –
That whatever Miss T. eats
Turns into Miss T.

W. DE LA MARE

Having cleared out of the way those objections which are based on a popular and confused notion that the 'progress of science' has somehow made the world safe against Miracle, we must now consider the subject on a somewhat deeper level. The question is whether Nature can be known to be of such a kind that supernatural interferences with her are impossible. She is already known to be, in general, regular: she behaves according to fixed laws, many of which have been discovered, and which interlock with one another. There is, in this discussion, no question of mere failure or inaccuracy to keep these laws on the part of

Nature, no question of chancy or spontaneous variation.[1] The only question is whether, granting the existence of a Power outside Nature, there is any intrinsic absurdity in the idea of its intervening to produce within Nature events which the regular 'going on' of the whole natural system would never have produced.

Three conceptions of the 'Laws' of Nature have been held. (1) That they are mere brute facts, known only by observation, with no discoverable rhyme or reason about them. We know *that* Nature behaves thus and thus; we do not know why she does and can see no reason why she should not do the opposite. (2) That they are applications of the law of averages. The foundations of Nature are in the random and lawless. But the number of units we are dealing with are so enormous that the behaviour of these crowds (like the behaviour of very large masses of men) can be calculated with practical accuracy. What we call 'impossible events' are events so overwhelmingly improbable—by actuarial standards—that we do not need to take them into account. (3) That the fundamental laws of Physics are really what we call 'necessary truths' like the truths of mathematics—in other words, that if we clearly under-

[1] If any region of reality is in fact chancy or lawless then it is a region which, so far from admitting Miracle with special ease, renders the word 'Miracle' meaningless throughout that region.

stand what we are saying we shall see that the opposite would be meaningless nonsense. Thus it is a 'law' that when one billiard ball shoves another the amount of momentum lost by the first ball must exactly equal the amount gained by the second. People who hold that the laws of Nature are necessary truths would say that all we have done is to split up the single events into two halves (adventures of ball A, and adventures of ball B) and then discover that 'the two sides of the account balance'. When we understand this we see that of course they *must* balance. The fundamental laws are in the long run merely statements that every event is itself and not some different event.

It will at once be clear that the first of these three theories gives no assurance against Miracles—indeed no assurance that, even apart from Miracles, the 'laws' which we have hitherto observed will be obeyed tomorrow. If we have no notion why a thing happens, then of course we know no reason why it should not be otherwise, and therefore have no certainty that it might not some day be otherwise. The second theory, which depends on the law of averages, is in the same position. The assurance it gives us is of the same general kind as our assurance that a coin tossed a thousand times will not give the same result, say, nine hundred times: and that the longer you toss it the more nearly the numbers of Heads and Tails will come to being equal. But this is so only provided the coin is an honest coin. If it is a loaded coin our expectations may be disappointed. But the people who believe in miracles are maintaining precisely that the coin *is* loaded. The expectations based on the law of averages will work only for *undoctored* Nature. And the question whether miracles occur is just the question whether Nature is ever doctored.

The third view (that laws of Nature are necessary truths) seems at first sight to present an insurmountable obstacle to miracle. The breaking of them would, in that case, be a self-contradiction and not even Omnipotence can do what is self-contradictory. Therefore the Laws cannot be broken. And therefore, we shall conclude, no miracle can ever occur?

We have gone too quickly. It is certain that the billiard balls will behave in a particular way, just as it is certain that if you divided a shilling unequally between two recipients then A's share must exceed the half and B's share fall short of it by exactly the same amount. Provided, of course, that A does not by sleight of hand steal some of B's pennies at the very moment of the transaction. In the same way, you know what will happen to the two billiard balls—provided nothing interferes. If one ball encounters a roughness in the cloth which the other does not, their motion will not illustrate the law in the way you had expected. Of course what happens as a result of the roughness in the cloth will illustrate the law in some other way, but your original prediction will have been false. Or again, if I snatch up a cue and give one of the balls a little help, you will get a third result: and that third result will equally illustrate the laws of physics, and equally falsify your prediction. I shall have 'spoiled the experiment'. All interferences leave the law perfectly true. But every prediction of what will happen in a given instance is made under the proviso 'other things being equal' or 'if there are no interferences'. Whether other things *are equal* in a given case and whether interferences may occur is another matter. The arithmetician, as an arithmetician, does not know how likely A is to steal some of B's pennies when the shilling is being divided; you had better ask a criminologist. The physicist, as a physicist, does not know how likely I am to catch up a cue and 'spoil' his experiment with the billiard balls: you had better ask someone who knows *me*. In the same way the physicist, as such, does not know how likely it is that some supernatural power is going to interfere with them: you had better ask a metaphysician. But the physicist does know, just because he is a physicist, that if the billiard balls are tampered with by any agency, natural or supernatural, which he has not taken into account, then their behaviour must differ from what he expected. Not because the law is false, but because it is true. The more certain we are of the law the more clearly we know that if new factors have been introduced the result will vary accordingly. What we do not know, as physicists, is whether Supernatural power might be one of the new factors.

If the laws of Nature are necessary truths, no miracle can break them: but then no miracle needs to break them. It is with them as with the laws of arithmetic. If I put six pennies into a drawer on Monday and six more on Tuesday, the laws decree that—*other things being equal*—I shall find twelve pennies there on Wednesday. But if the drawer has been robbed I may in fact find only two. Something will have been broken (the lock of the drawer or the laws of England) but the laws of arithmetic will not have been broken. The new situation created by the thief will illustrate the laws of arithmetic just as well as the original situation. But if God comes to work miracles, He comes 'like a thief in the night'. Miracle is, from the point of view of the scientist, a form of doctoring, tampering, (if you like) cheating. It introduces a new factor into the situation, namely supernatural force, which the scientist had not reckoned on. He calculates what will happen, or what must have happened on a past occasion, in the belief that the situation, at that point of space and time, is or was A. But if supernatural force has been added, then the situation really is or was AB. And no one knows better than the scientist that AB *cannot* yield the same result as A. The necessary truth of the laws, far from making it impossible that miracles should occur, makes it certain that if the Supernatural is operating they must occur. For if the natural situation by itself, and the natural situation *plus* something else, yielded only the same result, it would be then that we should be faced with a lawless and unsystematic universe. The better you know that two and two make four, the better you know that two and three don't.

This perhaps helps to make a little clearer what the laws of Nature really are. We are in the habit of talking as if they caused events to happen; but they have never caused any event at all. The laws of motion do not set billiard balls moving: they analyse the motion after something else (say, a man with a cue, or a lurch of the liner, or, perhaps, supernatural power) has provided it. They produce no events: they state the pattern to which every event—if only it can be induced to happen—must conform, just as the rules of arithmetic state the pattern to which all transactions with money must conform—if only you can get hold of any money. Thus in one sense the laws of Nature cover the whole field of space and time; in another, what they leave out is precisely the whole real universe—the incessant torrent of actual events which makes up true history. That must come from somewhere else. To think the laws can produce it is like thinking that you can create real money by simply doing sums. For every law, in the last resort, says 'If you have A, then you will get B'. But first catch your A: the laws won't do it for you.

It is therefore inaccurate to define a miracle as something that breaks the laws of Nature. It doesn't. If I knock out my pipe I alter the position of a great many atoms: in the long run, and to an infinitesimal degree, of all the atoms there are. Nature digests or assimilates this event with perfect ease and harmonises it in a twinkling with all other events. It is one more bit of raw material for the laws to apply to, and they apply. I have simply thrown one event into the general cataract of events and it finds itself at home there and conforms to all other events. If God annihilates or creates or deflects a unit of matter He has created a new situation at that point. Immediately all Nature domiciles this new situation, makes it at home in her realm, adapts all other events to it. It finds itself conforming to all the laws. If God creates a miraculous spermatozoon in the body of a virgin, it does not proceed to break any laws. The laws at once take it over. Nature is ready. Pregnancy follows, according to all the normal laws, and nine months later a child is born. We see every day that physical nature is not in the least incommoded by the daily inrush of events from biological nature or from psychological nature. If events ever come from beyond Nature altogether, she will be no more incommoded by them. Be sure she will rush to the point where she is invaded, as the defensive forces rush to a cut in our finger, and there hasten to accommodate the newcomer. The moment it enters her realm it obeys all her laws. Miraculous wine will intoxicate, miraculous conception will lead to pregnancy, inspired books will suffer all the ordinary processes of textual corruption, miraculous bread will be digested. The divine art of miracle is not an art of suspending the pattern to which events conform but of feeding new events into that pattern. It does not violate the law's proviso, 'If A, then B': it says, 'But this time instead of A, A2,' and Nature, speaking through all her laws, replies 'Then B2' and naturalises the immigrant, as she well knows how. She is an accomplished hostess.

A miracle is emphatically not an event without cause or without results. Its cause is the activity of God: its results follow according to Natural law. In the forward direction (i.e. during the time which follows its occurrence) it is interlocked with all Nature just like any other event. Its peculiarity is that it is not in that way interlocked backwards, interlocked with the previous history of Nature. And this is just what some people find intolerable. The reason they find it intolerable is that they start by taking Nature to be the whole of reality. And they are sure that all reality must be interrelated and consistent. I agree with them. But I think they have mistaken a partial system within reality, namely Nature, for the whole. That being so, the miracle and the previous history of Nature may be interlocked after all but not in the way the Naturalist expected: rather in a much more roundabout fashion. The great complex event called Nature, and the new particular event introduced into it by the miracle, are related by their common origin in God, and doubtless, if we knew enough, most intricately related in His purpose and design, so that a Nature which had had a different history, and therefore been a different Nature, would have been invaded by different miracles or by none at all. In that way the miracles and the previous course of Nature are as well interlocked as any other two realities, but you must go back as far as their common Creator to find the interlocking. You will not find it *within* Nature. The same sort of thing happens with any partial system. The behaviour of fishes which are being studied in a tank makes a relatively closed system. Now suppose that the tank is shaken by a bomb in the neighbourhood of the laboratory. The behaviour of the fishes will now be no longer fully explicable by what was going on in the tank before the bomb fell: there will be a failure of backward interlocking. This does not mean that the bomb and the previous history of events within the tank are totally and finally unrelated. It does mean that to find their relation you must go back to the much larger reality which includes both the tank and the bomb—the reality of wartime England in which bombs are falling but some laboratories are still at work. You would never find it within the history of the tank. In the same way, the miracle is not *naturally* interlocked in the backward direction. To find out how it is interlocked with the previous history of Nature you must replace both Nature and the miracle in a larger context. Everything *is* connected with everything else: but not all things are connected by the short and straight roads we expected.

The rightful demand that all reality should be consistent and systematic does not therefore exclude miracles: but it has a very valuable contribution to make to our conception of them. It reminds us that miracles, if they occur, must, like all events, be revelations of that total harmony of all that exists. Nothing arbitrary, nothing simply 'stuck on' and left unreconciled with the texture of total reality, can be admitted. By definition, miracles must of course interrupt the usual course of Nature; but if they are real they must, in the very act of so doing, assert all the more the unity and self-consistency of total reality at some deeper level. They will not be like unmetrical lumps of prose breaking the unity of a poem; they will be like that crowning metrical audacity which, though it may be paralleled nowhere else in the poem, yet, coming just where it does, and effecting just what it effects, is (to those who understand) the supreme revelation of the unity in the poet's conception. If what we call Nature is modified by supernatural power, then we may be sure that the capability of being so modified is of the essence of Nature—that the total events, if we could grasp it, would turn out to involve, by its very character, the possibility of such modifications. If Nature brings forth miracles then doubtless it is as 'natural' for her to do so when impregnated by the masculine force beyond her as it is for a woman to bear children to a man. In calling them miracles we do not mean that they are contradictions or outrages; we mean that, left to her own resources, she could never produce them.

9

A CHAPTER NOT STRICTLY NECESSARY

> And there we saw the giants, the sons of Anak; which come of the giants: and we were in our own sight as grasshoppers, and so we were in their sight
>
> *Numbers* 13:33

The last two chapters have been concerned with objections to Miracle, made, so to speak, from the side of Nature; made on the ground that she is the sort of system which could not admit miracles. Our next step, if we followed a strict order, would be to consider objections from the opposite side—in fact, to inquire whether what is beyond Nature can reasonably be supposed to be the sort of being that could, or would, work miracles. But I find myself strongly disposed to turn aside and face first an objection of a different sort. It is a purely emotional one; severer readers may skip this chapter. But I know it is one which weighed very heavily with me at a certain period of my life, and if others have passed through the same experience they may care to read of it.

One of the things that held me back from Supernaturalism was a deep repugnance to the view of Nature which, as I thought, Supernaturalism entailed. I passionately desired that Nature should exist 'on her own'. The idea that she had been made, and could be altered, by God, seemed to take from her all that spontaneity which I found so refreshing. In order to breathe freely I wanted to feel that in Nature one reached at last something that simply *was:* the thought that she had been manufactured or 'put there', and put there with a purpose, was suffocating. I wrote a poem in those days about a sunrise, I remember, in which, after describing the scene, I added that some people liked to believe there was a Spirit behind it all and that this Spirit was communicating with them. But, said I, that was exactly what I did not want. The poem was not much good and I have forgotten most of it: but it ended up by saying how much rather I would feel

That in their own right earth and sky
Continually do dance
For their own sakes—and here crept I
To watch the world by chance.

'*By chance!*'—one could not bear to feel that the sunrise had been in any way 'arranged' or had anything to do with oneself. To find that it had not simply happened, that it had been somehow contrived, would be as bad as finding that the fieldmouse I saw beside some lonely hedge was really a clockwork mouse put there to amuse me, or (worse still) to point some moral lesson. The Greek poet asks, 'If water sticks in your throat, what will you take to wash it down?' I likewise asked, 'If Nature herself proves artificial, where will you go to seek wildness? Where is the real out-of-doors?' To find that all the woods, and small streams in the middle of the woods, and odd corners of mountain valleys, and the wind and the grass were only a sort of *scenery*, only backcloths for some kind of play, and that play perhaps one with a moral—what flatness, what an anti-climax, what an unendurable bore!

The cure of this mood began years ago: but I must record that the cure was not complete until I began to study this question of Miracles. At every stage in the writing of this book I have found my idea of Nature becoming more vivid and more concrete. I set out on a work which seemed to involve reducing her status and undermining her walls at every turn: the paradoxical result is a growing sensation that if I am not very careful she will become the heroine of my book. She has never seemed to me more great or more real than at this moment.

The reason is not far to seek. As long as one is a Naturalist, 'Nature' is only a word for 'everything'. And Everything is not a subject about which anything very interesting can be said or (save by illusion) felt. One aspect of things strikes us and we talk of the 'peace' of Nature; another strikes us and we talk of her cruelty. And then, because we falsely take her for the ultimate and self-existent Fact and cannot quite repress our high instinct to worship the Self-existent, we are all at sea and our moods fluctuate and Nature means to us whatever we please as the moods select and slur. But everything becomes different when we recognise that Nature is a *creature,* a created thing, with its own particular tang or flavour. There is no need any longer to select and slur. It is not in her, but in Something far beyond her, that all lines meet and all contrasts are explained. It is no more baffling that the creature called Nature should be both fair and cruel than that the first man you meet in the train should be a dishonest grocer and a kind husband. For she is not the Absolute: she is one of the creatures, with her good points and her bad points and her own unmistakable flavour running through them all.

To say that God has created her is not to say that she is unreal, but precisely that she is real. Would you make God less creative than Shakespeare or Dickens? What He creates is created in the round: it is far more concrete than Falstaff or Sam Weller. The theologians certainly tell us that He created Nature freely. They mean that He was not forced to do so by any external necessity. But we must not interpret freedom negatively, as if Nature were a mere construction of parts arbitrarily stuck together. God's creative freedom is to be conceived as the freedom of a poet: the freedom to create a consistent, positive thing with its own inimitable flavour. Shakespeare need not create Falstaff: but if he does, Falstaff *must* be fat. God need not create this Nature. He might have created others, He may have created others. But granted *this* Nature, then doubtless no smallest part of her is there except because it expresses the character He chose to give her. It would be a miserable error to suppose that the dimensions of space and time, the death and rebirth of vegetation, the unity in multiplicity of organisms, the union in opposition of sexes, and the colour of each particular apple in Herefordshire this autumn, were merely a collection of useful devices forcibly welded together. They are the very idiom, almost the facial expression, the smell or taste, of an individual thing. The *quality* of Nature is present in them all just as the Latinity of Latin is present in every inflection or the 'Correggiosity' of Correggio in every stroke of the brush.

Nature is by human (and probably by Divine) standards partly good and partly evil. We Christians believe that she has been corrupted. But the same tang or flavour runs through both her corruptions and her excellences. Everything is in character. Falstaff does not sin in the same way as Othello. Othello's fall bears a close relation to his virtues. If Perdita had fallen she would not have been bad in the same way as Lady Macbeth: if Lady Macbeth had remained good her goodness would have been quite different from that of Perdita. The evils we see in Nature are, so to speak, the evils proper to *this* Nature. Her very character decreed that if she were corrupted the corruption would take this form and not another. The horrors of parasitism and the glories of motherhood are good and evil worked out of the same basic scheme or idea.

I spoke just now about the Latinity of Latin. It is more evident to us than it can have been to the Romans. The Englishness of English is audible only to those who know some other language as well. In the same way and for the same reason, only Supernaturalists really see Nature. You must go a little away from her, and then turn round, and look back. Then at last the true landscape will become visible. You must have tasted, however briefly, the pure water from beyond the world before you can be distinctly conscious of the hot, salty tang of Nature's current. To treat her as God, or as Everything, is to lose the whole pith and pleasure of her. Come out, look back, and then you will see . . . this astonishing cataract of bears, babies, and bananas: this immoderate deluge of atoms, orchids, oranges, cancers, canaries, fleas, gases, tornadoes and toads. How could you ever have thought this was the ultimate reality? How could you ever have thought that it was merely a stage-set for the moral drama of men and women? She is herself. Offer her neither worship nor contempt. Meet her and know her. If we are immortal, and if she is doomed (as the scientists tell us) to run down and die, we shall miss this half-shy and half-flamboyant creature, this ogress, this hoyden, this incorrigible fairy, this dumb witch. But the theologians tell us that she, like ourselves, is to be redeemed. The 'vanity' to which she was subjected was her disease, not her essence. She will be cured in character: not tamed (Heaven forbid) nor sterilised. We shall still be able to recognise our old enemy, friend, playfellow and foster-mother, so perfected as to be not less, but more, herself. And that will be a merry meeting.

10

'HORRID RED THINGS'

> We can call the attempt to refute theism by displaying the continuity of the belief in God with primitive delusion the method of Anthropological intimidation.
>
> EDWYN BEVAN, *Symbolism and Belief*, chap. ii.

I have argued that there is no security against Miracle to be found by the study of Nature. She is not the whole of reality but only a part; for all we know she might be a small part. If that which is outside her wishes to invade her she has, so far as we can see, no defences. But of course many who disbelieve in Miracles would admit all this. Their objection comes from the other side. They think that the Supernatural would not invade: they accuse those who say that it has done so of having a childish and unworthy notion of the Supernatural. They therefore reject all forms of Supernaturalism which assert such interference and invasions: and specially the form called Christianity, for in it the

Miracles, or at least some Miracles, are more closely bound up with the fabric of the whole belief than in any other. All the essentials of Hinduism would, I think, remain unimpaired if you subtracted the miraculous, and the same is almost true of Mohammedanism. But you cannot do that with Christianity. It is precisely the story of a great Miracle. A naturalistic Christianity leaves out all that is specifically Christian.

The difficulties of the unbeliever do not begin with questions about this or that particular miracle; they begin much further back. When a man who has had only the ordinary modern education looks into any authoritative statement of Christian doctrine, he finds himself face to face with what seems to him a wholly 'savage' or 'primitive' picture of the universe. He finds that God is supposed to have had a 'Son', just as if God were a mythological deity like Jupiter or Odin. He finds that this 'Son' is supposed to have 'come down from Heaven', just as if God had a palace in the sky from which He had sent down His 'Son' like a parachutist. He finds that this 'Son' then 'descended into Hell'—into some land of the dead under the surface of a (presumably) flat earth—and thence 'ascended' again, as if by a balloon, into his Father's sky-palace, where He finally sat down in a decorated chair placed a little to His Father's right. Everything seems to presuppose a conception of reality which the increase of our knowledge has been steadily refuting for the last two thousand years and which no honest man in his senses could return to today.

It is this impression which explains the contempt, and even disgust, felt by many people for the writings of modern Christians. When once a man is convinced that Christianity *in general* implies a local 'Heaven', a flat earth, and a God who can have children, he naturally listens with impatience to our solutions of particular difficulties and our defences against particular objections. The more ingenious we are in such solutions and defences the more perverse we seem to him. 'Of course,' he says, 'once the doctrines are there, clever people can invent clever arguments to defend them, just as, when once a historian has made a blunder he can go on inventing more and more elaborate theories to make it appear that it was not a blunder. But the real point is that none of these elaborate theories would have been thought of if he had read his documents correctly in the first instance. In the same way, is it not clear that Christian theology would never have come into existence at all if the writers of the New Testament had had the slightest knowledge of what the real universe is actually like?' Thus, at any rate, I used to think myself. The very man who taught me to think—a hard, satirical atheist (ex-Presbyterian) who doted on the *Golden Bough* and filled his house with the products of the Rationalist Press Association—thought in the same way; and he was a man as honest as the daylight, to whom I here willingly acknowledge an immense debt. His attitude to Christianity was for me the starting point of adult thinking; you may say it is bred in my bones. And yet, since those days, I have come to regard that attitude as a total misunderstanding.

Remembering, as I do, from within, the attitude of the impatient sceptic, I realise very well how he is fore-armed against anything I may say for the rest of this chapter. 'I know exactly what this man is going to do,' he murmurs. 'He is going to start explaining all these mythological statements away. It is the invariable practice of these Christians. On any matter whereon science has not yet spoken and on which they cannot be checked, they will tell you some preposterous fairytale. And then, the moment science makes a new advance and shows (as it invariably does) their statement to be untrue, they suddenly turn round and explain that they didn't mean what they said, that they were using a poetic metaphor or constructing an allegory, and that all they really intended was some harmless moral platitude. We are sick of this theological thimble-rigging'. Now I have a great deal of sympathy with that sickness and I freely admit that 'modernist' Christianity has constantly played just the game of which the impatient sceptic accuses it. But I also think there is a kind of explaining which is not explaining away. In one sense I am going to do just what the sceptic thinks I am going to do: that is, I am going to distinguish what I regard as the 'core' or 'real meaning' of the doctrines from that in their expression which I regard as inessential and possibly even capable of being changed without damage. But then, what will drop away from the 'real meaning' under my treatment will precisely *not* be the miraculous. It is the core itself, the core scraped as clean of inessentials as we can scrape it, which remains for me entirely miraculous, supernatural—nay, if you will, 'primitive' and even 'magical'.

In order to explain this I must now touch on a subject which has an importance quite apart from our present purpose and of which everyone who wishes to think clearly should make himself master as soon as he possibly can. And he ought to begin by reading Mr Owen Barfield's *Poetic Diction* and Mr Edwyn Bevan's *Symbolism and Belief*. But for the present argument it will be enough to leave the deeper problems on one side and proceed in a 'popular' and unambitious manner.

When I think about London I usually see a mental picture of Euston Station. But when I think (as I do) that

London has several million inhabitants, I do not mean that there are several million images of people contained in my image of Euston Station. Nor do I mean that several millions of real people live in the real Euston Station. In fact though I have the image while I am thinking about London, what I think or say is not *about* that image, and would be manifest nonsense if it were. It makes sense because it is not about my own mental pictures but about the real London, outside my imagination, of which no one can have an adequate mental picture at all. Or again, when we say that the Sun is ninety-odd million miles away, we understand perfectly clearly what we mean by this number; we can divide and multiply it by other numbers and we can work out how long it would take to travel that distance at any given speed. But this clear *thinking* is accompanied by *imagining* which is ludicrously false to what we know that the reality must be.

To think, then, is one thing, and to imagine is another. What we think or say can be, and usually is, quite different from what we imagine or picture; and what we mean may be true when the mental images that accompany it are entirely false. It is, indeed, doubtful whether anyone except an extreme visualist who is also a trained artist ever has mental images which are particularly like the things he is thinking about.

In these examples the mental image is not only unlike the reality but is known to be unlike it, at least after a moment's reflection. I know that London is not merely Euston Station. Let us now go on to a slightly different predicament. I once heard a lady tell her young daughter that you would die if you ate too many tablets of aspirin. 'But why?' asked the child, 'it isn't poisonous'. 'How do you know it isn't poisonous?' said the mother. 'Because', said the child, 'when you crush an aspirin tablet you don't find horrid red things inside it'. Clearly, when this child thought of poison she had a mental picture of Horrid Red Things, just as I have a picture of Euston when I think of London. The difference is that whereas I know my image to be very unlike the real London, the child thought that poison was *really red.* To that extent she was mistaken. But this does not mean that everything she thought or said about poison was necessarily nonsensical. She knew perfectly well that a poison was something which killed you or made you ill if you swallowed it; and she knew, to some extent, which of the substances in her mother's house were poisonous. If a visitor to that house had been warned by the child, 'Don't drink that. Mother says it is poison', he would have been ill advised to neglect the warning on the ground that 'This child has a primitive idea of poison as Horrid Red Things, which my adult scientific knowledge has long since refuted.'

We can now add to our previous statement (that thinking may be sound where the images that accompany it are false) the further statement: thinking may be sound in certain respects where it is accompanied not only by false images but by false images mistaken for true ones.

There is still a third situation to be dealt with. In our two previous examples we have been concerned with thought and imagination, but not with language. I had to picture Euston Station, but I did not need to *mention* it; the child thought that poison was Horrid Red Things, but she could talk about poison without saying so. But very often when we are talking about something which is not perceptible by the five senses we use words which, in one of their meanings, refer to things or actions that are. When a man says that he grasps an argument he is using a verb (*grasp*) which literally means to take something in the hands, but he is certainly not thinking that his mind has hands or that an argument can be seized like a gun. To avoid the word *grasp* he may change the form of expression and say, 'I see your point,' but he does not mean that a pointed object has appeared in his visual field. He may have a third shot and say, 'I follow you,' but he does not mean that he is walking behind you along a road. Everyone is familiar with this linguistic phenomenon and the grammarians call it metaphor. But it is a serious mistake to think that metaphor is an optional thing which poets and orators may put into their work as a decoration and plain speakers can do without. The truth is that if we are going to talk at all about things which are not perceived by the senses, we are forced to use language metaphorically. Books on psychology or economics or politics are as continuously metaphorical as books of poetry or devotion. There is no other way of talking, as every philologist is aware. Those who wish can satisfy themselves on the point by reading the books I have already mentioned and the other books to which those two will lead them on. It is a study for a lifetime and I must here content myself with the mere statement; all speech about supersensibles is, and must be, metaphorical in the highest degree.

We have now three guiding principles before us. (1) That thought is distinct from the imagination which accompanies it. (2) That thought may be in the main sound even when the false images that accompany it are mistaken by the thinker for true ones. (3) That anyone who talks about things that cannot be seen, or touched, or heard, or the like, must inevitably talk as *if they could be* seen or touched or heard (e.g. must talk of 'complexes' and 'repressions' *as if* desires could really be tied up in bundles or shoved back; of 'growth' and 'development' *as if* institutions could really grow like trees or unfold like flowers; of energy being 'released' *as if* it were an animal let out of a cage).

Let us now apply this to the 'savage' or 'primitive' articles of the Christian creed. And let us admit at once that many Christians (though by no means all) when they make these assertions do have in mind just those crude mental pictures which so horrify the sceptic. When they say that Christ 'came down from Heaven' they do have a vague image of something shooting or floating downwards out of the sky. When they say that Christ is the 'Son' of 'the Father' they may have a picture of two human forms, the one looking rather older than the other. But we now know that the mere presence of these mental pictures does not, of itself, tell us anything about the reasonableness or absurdity of the thoughts they accompany. If absurd images meant absurd thought, then we should all be thinking nonsense all the time. And the Christians themselves make it clear that the images are not to be identified with the thing believed. They may picture the Father as a human form, but they also maintain that He has no body. They may picture Him older than the son, but they also maintain the one did not exist before the other, both having existed from all eternity. I am speaking, of course, about Christian adults. Christianity is not to be judged from the fancies of children any more than medicine from the ideas of the little girl who believed in horrid red things.

At this stage I must turn aside to deal with a rather simpleminded illusion. When we point out that what the Christians mean is not to be identified with their mental pictures, some people say, 'In that case, would it not be better to get rid of the mental pictures, and of the language which suggests them, altogether?' But this is impossible. The people who recommend it have not noticed that when they try to get rid of man-like, or as they are called, 'anthropomorphic', images they merely succeed in substituting images of some other kind. 'I don't believe in a personal God,' says one, 'but I do believe in a great spiritual force'. What he has not noticed is that the word 'force' has let in all sorts of images about winds and tides and electricity and gravitation. 'I don't believe in a personal God,' says another, 'but I do believe we are all parts of one great Being which moves and works through us all'—not noticing that he has merely exchanged the image of a fatherly and royal-looking man for the image of some widely extended gas or fluid. A girl I knew was brought up by 'higher thinking' parents to regard God as a perfect 'substance'; in later life she realised that this had actually led her to think of Him as something like a vast tapioca pudding. (To make matters worse, she disliked tapioca). We may feel ourselves quite safe from this degree of absurdity, but we are mistaken. If a man watches his own mind, I believe he will find that what profess to be specially advanced or philosophic conceptions of God are, in his thinking, always accompanied by vague images which, if inspected, would turn out to be even more absurd than the man-like images aroused by Christian theology. For man, after all, is the highest of the things we meet in sensuous experience. He has, at least, conquered the globe, honoured (though not followed) virtue, achieved knowledge, made poetry, music and art. If God exists at all it is not unreasonable to suppose that we are less unlike Him than anything else we know. No doubt we are unspeakably different from Him; to that extent all man-like images are false. But those images of shapeless mists and irrational forces which, unacknowledged, haunt the mind when we think we are rising to the conception of impersonal and absolute Being, must be very much more so. For images, of the one kind or of the other, will come; we cannot jump off our own shadow.

As far, then, as the adult Christian of modern times is concerned, the absurdity of the images does not imply absurdity in the doctrines; but it may be asked whether the early Christian was in the same position. Perhaps he mistook the images for true ones, and really believed in the sky-palace or the decorated chair. But as we have seen from the example of the Horrid Red Things, even this would not necessarily invalidate everything that he thought on these subjects. The child in our example might know many truths about poison and even, in some particular cases, truths which a given adult might not know. We can suppose a Galilean peasant who thought that Christ had literally and physically 'sat down at the right hand of the Father'. If such a man had then gone to Alexandria and had a philosophical education he would have discovered that the Father had no right hand and did not sit on a throne. Is it conceivable that he would regard this as making any difference to what he had really intended and valued, in the doctrine during the days of his naïvety? For unless we suppose him to have been not only a peasant but a fool (two very different things) physical details about a supposed celestial throne-room would not have been what he cared about. What mattered must have been the belief that a person whom he had known as a man in Palestine had, as a person, survived death and was now operating as the supreme agent of the supernatural Being who governed and maintained the whole field of reality. And that belief would survive substantially unchanged after the falsity of the earlier images had been recognised.

Even if it could be shown, then, that the early Christians accepted their imagery literally, this would not mean that we are justified in relegating their doctrines as a whole to the lumber-room. Whether they actually did, is another matter. The difficulty here is that they were not writing as philosophers to satisfy speculative curiosity about the nature of God and of the universe. They *believed* in God; and once a man does that, philosophical definiteness can never be the *first* necessity. A drowning man does not analyse the rope that is flung at him, nor an impassioned lover consider the chemistry of his mistress's complexion. Hence the sort of question we are now considering is never raised by the New Testament writers. When once it is raised, Christianity decides quite clearly that the naïf images are false. The sect in the Egyptian desert which thought that God was like a man is condemned: the desert monk who felt he had lost something by its correction is recognised as 'muddle-headed'.[1] All three Persons of the Trinity are declared 'incomprehensible'.[2] God is pronounced 'inexpressible, unthinkable, invisible to all created beings'.[3] The Second Person is not only bodiless but so unlike man that if self-revelation had been His sole purpose He would not have chosen to be incarnate in a human form.[4] We do not find similar statements in the New Testament, because the

---

[1] *Senex mente confusus* Cassian quoted in Gibbon, cap. xlvii.

[2] Athanasian Creed.

[3] St Chrysostom *De Incomprehensibili* quoted in Otto, *Idea of the Holy*, Appendix 1.

[4] Athanasius De Incarnatione viii.

issue has not yet been made explicit: but we do find statements which make it certain how that issue will be decided when once it becomes explicit. The title 'Son' may sound 'primitive' or 'naïf'. But already in the New Testament this 'Son' is identified with the Discourse or Reason or Word which was eternally 'with God' and yet also *was* God.[5] He is the all-pervasive principle of concretion or cohesion whereby the universe holds together.[6] All things, and specially Life, arose *within* Him,[7] and within Him all things will reach their conclusion—the final statement of what they have been trying to express.[8]

It is, of course, always possible to imagine an earlier stratum of Christianity from which such ideas were absent; just as it is always possible to say that anything you dislike in Shakespeare was put in by an 'adapter' and the original play was free from it. But what have such assumptions to do with serious inquiry? And here the fabrication of them is specially perverse, since even if we go back beyond Christianity into Judaism itself, we shall not find the unambiguous anthropomorphism (or man-likeness) we

---

[5] John 1:1.

[6] Colossians 1:17.

[7] Colossians 1 ἐν αὐτῳ ἐχτισθη. John 1:4.

[8] Ephesians 1:10.

are looking for. Neither, I admit, shall we find its denial. We shall find, on the one hand, God pictured as living above 'in the high and holy place': we shall find, on the other, 'Do not I fill heaven and earth? saith the Lord'.[9] We shall find that in Ezekiel's vision God appeared (notice the hesitating words) in 'the likeness as the appearance of a man'.[10] But we shall find also the warning, 'Take ye therefore good heed unto yourselves. For ye saw no manner of similitude on the day that the Lord spake unto you in Horeb out of the midst of the fire—lest ye corrupt yourselves and make a graven image'.[11] Most baffling of all to a modern literalist, the God who seems to live locally in the sky, also *made* it.[12]

The reason why the modern literalist is puzzled is that he is trying to get out of the old writers something which is not there. Starting from a clear modern distinction between material and immaterial he tries to find out on which side of that distinction the ancient Hebrew conception fell. He forgets that the distinction itself has been made clear only by later thought.

We are often told that primitive man could not conceive

---

[9] Jeremiah 23:24.

[10] Ezekiel 1:26.

[11] Deuteronomy 4:15.

[12] Genesis 1:1.

pure spirit; but then neither could he conceive mere matter. A throne and a local habitation are attributed to God only at that stage when it is still impossible to regard the throne, or palace even of an earthly king as merely physical objects. In earthly thrones and palaces it was the spiritual significance—as we should say, the 'atmosphere'—that mattered to the ancient mind. As soon as the contrast of 'spiritual' and 'material' was before their minds, they knew God to be 'spiritual' and realised that their religion had implied this all along. But at an earlier stage that contrast was not there. To regard that earlier stage as unspiritual because we find there no clear assertion of unembodied spirit, is a real misunderstanding. You might just as well call it spiritual because it contained no clear consciousness of mere matter. Mr Barfield has shown, as regards the history of language, that words did not start by referring merely to physical objects and then get extended by metaphor to refer to emotions, mental states and the like. On the contrary, what we now call the 'literal and metaphorical' meanings have both been disengaged by analysis from an ancient unity of meaning which was neither or both. In the same way it is quite erroneous to think that man started with a 'material' God or 'Heaven' and gradually spiritualised them. He could not have started with something 'material' for the 'material', as we understand it, comes to be realised only by contrast to the 'immaterial', and the two sides of the contrast grow at the same speed. He started with something which was neither and both. As long as we are trying to read back into that ancient unity either the one or the other of the two opposites which have since been analysed out of it, we shall misread all early literature and ignore many states of consciousness which we ourselves still from time to time experience. The point is crucial not only for the present discussion but for any sound literary criticism or philosophy.

The Christian doctrines, and even the Jewish doctrines which preceded them, have always been statements about spiritual reality, not specimens of primitive physical science. Whatever is positive in the conception of the spiritual has always been contained in them; it is only its negative aspect (immateriality) which has had to wait for recognition until abstract thought was fully developed. The material imagery has never been taken literally by anyone who had reached the stage when he could understand what 'taking it literally' meant. And now we come to the difference between 'explaining' and 'explaining away'. It shows itself in two ways. (1) Some people when they say that a thing is meant 'metaphorically' conclude from this that it is hardly meant at all. They rightly think that Christ spoke metaphorically when he told us to carry the cross: they wrongly conclude that carrying the cross means nothing more than leading a respectable life and subscribing moderately to charities. They reasonably think that hell 'fire' is a metaphor—and unwisely conclude that it means nothing more serious than remorse. They say that the story of the Fall in Genesis is not literal; and then go on to say (I have heard them myself) that it was really a fall upwards—which is like saying that because 'My heart is broken' contains a metaphor, it therefore means 'I feel very cheerful'. This mode of interpretation I regard, frankly, as nonsense. For me the Christian doctrines which are 'metaphorical'—or which have become metaphorical with the increase of abstract thought—mean something which is just as 'supernatural' or shocking after we have removed the ancient imagery as it was before. They mean that in addition to the physical or psycho-physical universe known to the sciences, there exists an uncreated and unconditioned reality which causes the universe to be; that this reality has a positive structure or constitution which is usefully, though doubtless not completely, described in the doctrine of the Trinity; and that this reality, at a definite point in time, entered the universe we know by becoming one of its own creatures and there produced effects on the historical level which the normal workings of the natural universe do not produce; and that this has brought about a change in our relations to the unconditioned reality. It will be noticed that our colourless 'entered the universe' is not a whit less metaphorical than the more picturesque 'came down from Heaven'. We have only substituted a picture of horizontal or unspecified movement for one of vertical movement. And every attempt to improve the ancient language will have the same result. These things not only cannot be asserted—they cannot even be presented for discussion—without metaphor. We can make our speech duller; we cannot make it more literal. (2) These statements concern two things—the supernatural, unconditioned reality, and those events on the historical level which its irruption into the natural universe is held to have produced. The first thing is indescribable in 'literal' speech, and therefore we rightly interpret all that is said about it metaphorically. But the second thing is in a wholly different position. Events on the historical level are the sort of things we can talk about literally. If they occurred, they were perceived by the senses of men. Legitimate 'explanation' degenerates into muddled or dishonest 'explaining away' as soon as we start applying to these events the metaphorical interpretation which we rightly apply to the statements about God. The assertion that God has a Son was never intended to mean that He is a being propagating His kind by sexual intercourse: and so we do not alter Christianity by rendering explicit the fact that 'sonship' is not used of Christ in exactly the same sense in which it is used of men. But the assertion that Jesus turned water into wine was meant perfectly literally, for this refers to something which, if it happened, was well within the reach of our senses and our language. When I say, 'My heart is broken,' you know perfectly well that I don't mean anything you could verify at a post-mortem. But when I say, 'My bootlace is broken,' then, if your own observation shows it to be intact, I am either lying or mistaken. The accounts of the 'miracles' in first-century Palestine are either lies, or legends, or history. And if all, or the most important, of them are lies or legends then the claim which Christianity has been making for the last two thousand years is simply false. No doubt it might even so contain noble sentiments and moral truths. So does Greek mythology; so does Norse. But that is quite a different affair.

Nothing in this chapter helps us to a decision about the probability or improbability of the Christian claim. We have merely removed a misunderstanding in order to secure for that question a fair hearing.

11

CHRISTIANITY AND 'RELIGION'

> Those who make religion their god will not have God for their religion.
>
> THOMAS ERSKINE OF LINLATHEN

Having eliminated the confusions which come from ignoring the relations of thought, imagination, and speech, we may now return to our question. The Christians say that God has done miracles. The modern world, even when it believes in God, and even when it has seen the defencelessness of Nature, does not. It thinks God would not do that sort of thing. Have we any reason for supposing that the modern world is right? I agree that the sort of God conceived by the popular 'religion' of our own times would almost certainly work no miracles. The question is whether that popular religion is at all likely to be true.

I call it 'religion' advisedly. We who defend Christianity find ourselves constantly opposed not by the irreligion of our hearers but by their real religion. Speak about beauty, truth and goodness, or about a God who is simply the indwelling principle of these three, speak about a great spiritual force pervading all things, a common mind of which we are all parts, a pool of generalised spirituality to which we can all flow, and you will command friendly interest. But the temperature drops as soon as you mention a God who has purposes and performs particular actions, who does one thing and not another, a concrete, choosing, commanding, prohibiting God with a determinate character. People become embarrassed or angry. Such a conception seems to them primitive and crude and even irreverent. The popular 'religion' excludes miracles because it excludes the 'living God' of Christianity and believes instead in a kind of God who obviously would not do miracles, or indeed anything else. This popular 'religion' may roughly be called Pantheism, and we must now examine its credentials.

In the first place it is usually based on a quite fanciful picture of the history of religion. According to this picture, Man starts by inventing 'spirits' to explain natural phenomena; and at first he imagines these spirits to be exactly like himself. As he gets more enlightened they become less man-like, less 'anthropomorphic' as the scholars call it. Their anthropomorphic attributes drop off one by one—first the human shape, the human passions, the personality, will, activity—in the end every concrete or positive attribute whatever. There is left in the end a pure abstraction—mind as such, spirituality as such. God, instead of being a particular entity with a real character of its own, becomes simply 'the whole show' looked at in a particular way or the theoretical point at which all the lines of human aspiration would meet if produced to infinity. And since, on the modern view, the final stage of anything is the most refined and civilised stage, this 'religion' is held to be a more profound, more spiritual, and more enlightened belief than Christianity.

Now this imagined history of religion is not true. Pantheism certainly is (as its advocates would say) congenial to the modern mind; but the fact that a shoe slips on easily does not prove that it is a new shoe—much less that it will keep your feet dry. Pantheism is congenial to our minds not because it is the final stage in a slow process of enlightenment, but because it is almost as old as we are. It may even be the most primitive of all religions, and the *orenda* of a savage tribe has been interpreted by some to be an 'all-pervasive spirit'. It is immemorial in India. The Greeks rose above it only at their peak, in the thought of Plato and Aristotle; their successors relapsed into the great Pantheistic system of the Stoics. Modern Europe escaped it only while she remained predominantly Christian; with

Giordano Bruno and Spinoza it returned. With Hegel it became almost the agreed philosophy of highly educated people, while the more popular Pantheism of Wordsworth, Carlyle and Emerson conveyed the same doctrine to those on a slightly lower cultural level. So far from being the final religious refinement, Pantheism is in fact the permanent natural bent of the human mind; the permanent ordinary level below which man sometimes sinks, under the influence of priestcraft and superstition, but above which his own unaided efforts can never raise him for very long. Platonism and Judaism, and Christianity (which has incorporated both) have proved the only things capable of resisting it. It is the attitude into which the human mind automatically falls when left to itself. No wonder we find it congenial. If 'religion' means simply what man says about God, and not what God does about man, then Pantheism almost *is* religion. And 'religion' in that sense has, in the long run, only one really formidable opponent—namely Christianity.[1] Modern philosophy has rejected Hegel and modern science started out with no bias in favour of reli-

[1] Hence, if a Minister of Education professes to value religion and at the same time takes steps to suppress Christianity, it does not necessarily follow that he is a hypocrite or even (in the ordinary this-wordly sense of the word) a fool. He may sincerely desire more 'religion' and rightly see that the suppression of Christianity is a necessary preliminary to his design.

gion; but they have both proved quite powerless to curb the human impulse toward Pantheism. It is nearly as strong today as it was in ancient India or in ancient Rome. Theosophy and the worship of the life-force are both forms of it: even the German worship of a racial spirit is only Pantheism truncated or whittled down to suit barbarians. Yet, by a strange irony, each new relapse into this immemorial 'religion' is hailed as the last word in novelty and emancipation.

This native bent of the mind can be paralleled in quite a different field of thought. Men believed in atoms centuries before they had any experimental evidence of their existence. It was apparently natural to do so. And the sort of atoms we naturally believe in are little hard pellets—just like the hard substances we meet in experience, but too small to see. The mind reaches this conception by an easy analogy from grains of sand or of salt. It explains a number of phenomena; and we feel at home with atoms of that sort—we can picture them. The belief would have lasted forever if later science had not been so troublesome as to find out what atoms are *really* like. The moment it does that, all our mental comfort, all the immediate plausibility and obviousness of the old atomic theory, is destroyed. The real atoms turn out to be quite alien from our natural mode of thought. They are not even made of hard 'stuff' or 'matter' (as the imagination understands 'matter') at all: they are not simple, but have a structure: they are not all the same: and they are unpicturable. The old atomic theory is in physics what Pantheism is in religion—the normal, instinctive guess of the human mind, not utterly wrong, but needing correction. Christian theology, and quantum physics, are both, by comparison with the first guess, hard, complex, dry and repellent. The first shock of the object's real nature, breaking in on our spontaneous dreams of what that object ought to be, always has these characteristics. You must not expect Schrödinger to be as plausible as Democritus; he knows too much. You must not expect St Athanasius to be as plausible as Mr Bernard Shaw: he also knows too much.

The true state of the question is often misunderstood because people compare an adult knowledge of Pantheism with a knowledge of Christianity which they acquired in their childhood. They thus get the impression that Christianity gives the 'obvious' account of God, the one that is too easy to be true, while Pantheism offers something sublime and mysterious. In reality, it is the other way round. The apparent profundity of Pantheism thinly veils a mass of spontaneous picture-thinking and owes its plausibility to that fact. Pantheists and Christians agree that God is present everywhere. Pantheists conclude that He is 'diffused' or 'concealed' in all things and therefore a universal medium rather that a concrete entity, because their minds are really dominated by the picture of a gas, or fluid, or space itself. The Christian, on the other hand, deliberately rules out such images by saying that God is totally present at every point of space and time, and *locally* present in none. Again the Pantheist and Christian agree that we are all dependent on God and intimately related to Him. But the Christian defines this relation in terms of Maker and made, whereas the Pantheist (at least of the popular kind) says, we are 'parts' of Him, or are contained in Him. Once more, the picture of a vast extended something which can be divided into areas has crept in. Because of this fatal picture Pantheism concludes that God must be equally present in what we call evil and what we call good and therefore indifferent to both (ether permeates the mud and the marble impartially). The Christian has to reply that this is far too simple; God is present in a great many different modes: not present in matter as He is present in man, not present in all men as in some, not present in any other man as in Jesus. Pantheist and Christian also agree that God is super-personal. The Christian means by this that God has a positive structure which we could never have guessed in advance, any more than a knowledge of squares would have enabled us to guess at a cube. He contains 'persons' (three of them) while remaining one God, as a cube contains six squares while remaining one solid body. We cannot comprehend such a structure any more than the Flatlanders could comprehend a cube. But we can at least comprehend our incomprehension, and see that if there is something beyond personality it *ought* to be incomprehensible in that sort of way. The Pantheist, on the other hand, though he may say 'super-personal' really conceives God in terms of what is sub-personal—as though the Flatlanders thought a cube existed in *fewer* dimensions than a square.

At every point Christianity has to correct the natural expectations of the Pantheist and offer something more difficult, just as Schrödinger has to correct Democritus. At every moment he has to multiply distinctions and rule out false analogies. He has to substitute the mappings of something that has a positive, concrete, and highly articulated character for the formless generalities in which Pantheism is at home. Indeed, after the discussion has been going on for some time, the Pantheist is apt to change his ground and where he before accused us of childish naïvety now to blame us for the pedantic complexity of our 'cold Christs and tangled Trinities'. And we may well sympathise with him. Christianity, faced with popular 'religion' is continuously troublesome. To the large well-meant statements of 'religion' it finds itself forced to reply again and again,

'Well, not quite like that,' or 'I should hardly put it that way'. This troublesomeness does not of course prove it to be true; but if it were true it would be bound to have this troublesomeness. The real musician is similarly troublesome to a man who wishes to indulge in untaught 'musical appreciation'; the real historian is similarly a nuisance when we want to romance about 'the old days' or 'the ancient Greeks and Romans'. The ascertained nature of any real thing is always at first a nuisance to our natural fantasies—a wretched, pedantic, logic-chopping intruder upon a conversation which was getting on famously without it.

But 'religion' also claims to base itself on experience. The experiences of the mystics (that ill-defined but popular class) are held to indicate that God is the God of 'religion' rather than of Christianity; that He—or It—is not a concrete Being but 'being in general' about which nothing can be truly asserted. To everything which we try to say about Him, the mystics tend to reply, 'Not thus'. What all these negatives of the mystics really mean I shall consider in a moment: but I must first point out why it seems to me impossible that they should be true in the sense popularly understood.

It will be agreed that, however they came there, concrete, individual, determinate things do now exist: things like flamingoes, German generals, lovers, sandwiches, pineapples, comets and kangaroos. These are not mere principles or generalities or theorems, but things—facts—real, resistant existences. One might even say *opaque* existences, in the sense that each contains something which our intelligence cannot completely digest. In so far as they illustrate general laws it can digest them: but then they are never mere illustrations. Above and beyond that there is in each of them the 'opaque' brute fact of existence, the fact that it is actually there and is itself. Now this opaque fact, this concreteness, is not in the least accounted for by the laws of Nature or even by the laws of thought. Every law can be reduced to the form 'If A, then B.' Laws give us only a universe of 'Ifs and Ands': not this universe which actually exists. What we know through laws and general principles is a series of connections. But in order for there to be a real universe the connections must be given something to connect; a torrent of opaque actualities must be fed into the pattern. If God created the world, then He is precisely the source of this torrent, and it alone gives our truest principles anything to be true *about*. But if God is the ultimate source of all concrete, individual things and events, then God Himself must be concrete, and individual in the highest degree. Unless the origin of all other things were itself concrete and individual, nothing else could be so; for there is no conceivable means whereby what is abstract or general could itself produce concrete reality. Bookkeeping, continued to all eternity, could never produce one farthing. Metre, of itself, could never produce a poem. Bookkeeping needs something else (namely, real money put into the account) and metre needs something else (real words, fed into it by a poet) before any income or any poem can exist. If anything is to exist at all, then the Original Thing must be, not a principle nor a generality, much less an 'ideal' or a 'value', but an utterly concrete fact.

Probably no thinking person would, in so many words, deny that God is concrete and individual. But not all thinking people, and certainly not all who believe in 'religion', keep this truth steadily before their minds. We must beware, as Professor Whitehead says, of paying God ill-judged 'metaphysical compliments'. We say that God is 'infinite'. In the sense that His knowledge and power extend not to some things but to all, this is true. But if by using the word 'infinite' we encourage ourselves to think of Him as a formless 'everything' about whom nothing in particular and everything in general is true, then it would be better to drop that word altogether. Let us dare to say that God is a particular Thing. Once He was the only Thing: but He is creative. He made other things to be. He is not those other things. He is not 'universal being': if He were there would be no creatures, for a generality can make nothing. He is 'absolute being'—or rather *the* Absolute Being—in the sense that He alone exists in His own right. But there are things which God is not. In that sense He has a determinate character. Thus He is righteous, not amoral; creative, not inert. The Hebrew writings here observe an admirable balance. Once God says simply I AM, proclaiming the mystery of self-existence. But times without number He says 'I am the Lord'—I, the ultimate Fact, have *this* determinate character, and not *that.* And men are exhorted to ' know the Lord', to discover and experience this particular character.

The error which I am here trying to correct is one of the most sincere and respectable errors in the world; I have sympathy enough with it to feel shocked at the language I have been driven to use in stating the opposite view, which I believe to be the true one. To say that God 'is a particular Thing' does seem to obliterate the immeasurable difference not only between what He is and what all other things are but between the very mode of His existence and theirs. I must at once restore the balance by insisting that derivative things, from atoms to archangels, hardly attain to existence at all in comparison with their Creator. Their principle of existence is not in themselves. You can distinguish *what* they are from the fact *that* they are. The definition of them can be understood and a clear idea of them formed without even knowing *whether* they are. Existence is an 'opaque' addition to the idea of them. But with God it is not so: if we fully understood *what* God is we should see that there is no question *whether* He is. It would always have been impossible that He should not exist. He is the opaque centre of all existences, the thing that simply and entirely *is,* the fountain of facthood. And yet, now that He has created, there is a sense in which we must say that He is a particular Thing and even one Thing among others. To say this is not to lessen the immeasurable difference between Him and them. On the contrary, it is to recognise in Him a positive perfection which Pantheism has obscured; the perfection of being creative. He is so brim-full of existence that He can give existence away, can cause things to be, and to be really other than Himself, can make it untrue to say that He is everything.

It is clear that there never was a time when nothing existed; otherwise nothing would exist now. But to exist means to be a positive Something, to have (metaphorically) a certain shape or structure, to be this and not that. The Thing which always existed, namely God, has therefore always had His own positive character. Throughout all eternity certain statements about Him would have been true and others false. And from the mere fact of our own existence and Nature's we already know to some extent which are which. We know that He invents, acts, creates. After that there can be no ground for assuming in advance that He does not do miracles.

Why, then, do the mystics talk of Him as they do, and why are many people prepared in advance to maintain that, whatever else God may be, He is not the concrete, living, willing, and acting God of Christian theology? I think the reason is as follows. Let us suppose a mystical limpet, a sage among limpets, who (rapt in vision) catches a glimpse of what Man is like. In reporting it to his disciples, who have some vision themselves (though less than he) he will have to use many negatives. He will have to tell them that Man has no shell, is not attached to a rock, is not surrounded by water. And his disciples, having a little vision of their own to help them, do get some idea of Man. But then there come erudite limpets, limpets who write histories of philosophy and give lectures on comparative religion, and who have never had any vision of their own. What they get out of the prophetic limpet's words is simply and solely the negatives. From these, uncorrected by any positive insight, they build up a picture of Man as a sort of amorphous jelly (he has no shell) existing nowhere in particular (he is not attached to a rock) and never taking nourishment (there is no water to drift it towards him). And having a traditional reverence for Man they conclude that to be a famished jelly in a dimensionless void is the supreme mode of existence, and reject as crude, materialistic superstition any doctrine which would attribute to Man a definite shape, a structure, and organs.

Our own situation is much like that of the erudite limpets. Great prophets and saints have an intuition of God which is positive and concrete in the highest degree. Because, just touching the fringes of His being, they have seen that He is plenitude of life and energy and joy, therefore (and for no other reason) they have to pronounce that He transcends those limitations which we call personality, passion, change, materiality, and the like. The positive quality in Him which repels these limitations is their only ground for all the negatives. But when we come limping after and try to construct an intellectual or 'enlightened' religion, we take over these negatives (infinite, immaterial, impassible, immutable, etc.) and use them unchecked by any positive intuition. At each step we have to strip off from our idea of God some human attribute. But the only real reason for stripping off the human attribute is to make room for putting in some positive divine attribute. In St Paul's language, the purpose of all this unclothing is not that our idea of God should reach nakedness but that it should be reclothed. But unhappily we have no means of doing the reclothing. When we have removed from our idea of God some puny human characteristic, we (as merely erudite or intelligent enquirers) have no resources from which to supply that blindingly real and concrete attribute of Deity which ought to replace it. Thus at each step in the process of refinement our idea of God contains less, and the fatal pictures come in (an endless, silent sea, an empty sky beyond all stars, a dome of white radiance) and we reach at last mere zero and worship a nonentity. And the understanding, left to itself, can hardly help following this path. That is why the Christian statement that only He who does the will of the Father will ever know the true doctrine is philosophically accurate. Imagination may help a little: but in the moral life, and (still more) in the devotional life we touch something concrete which will at once begin to correct the growing emptiness of our idea of God. One moment even of feeble contrition or blurred thankfulness will, at least in some degree, head us off from the abyss of abstraction. It is Reason herself which teaches us not to rely on Reason only in this matter. For Reason knows that she cannot work without materials. When it becomes clear that you cannot find out by reasoning whether the cat is in the linen-cupboard, it is Reason herself who whispers, 'Go and look. This is not my job: it is a matter for the senses'. So here. The materials for correcting our abstract conception of God cannot be supplied by Reason: she will be the first to tell you to go and try experience—'Oh, taste and see!' For of course she will have already pointed out that your present position is absurd. As long as we remain Erudite Limpets we are forgetting that if no one had ever seen more of God than we, we should have no reason even to believe Him immaterial, immutable, impassible and all the rest of it. Even that negative knowledge which seems to us so enlightened is only a relic left over from the positive knowledge of better men—only the pattern which that heavenly wave left on the sand when it retreated.

'A Spirit and a Vision,' said Blake, 'are not, as the modern philosophy supposes, a cloudy vapour, or a nothing. They are organised and minutely articulated beyond all that the mortal and perishing nature can produce'.[2] He is speaking only of how to draw pictures of apparitions which may well have been illusory, but his words suggest a truth on the metaphysical level also. God is basic Fact or Actuality, the source of all other facthood. At all costs therefore He must not be thought of as a featureless generality. If He exists at all, He is the most concrete thing there is, the most individual, 'organised and minutely articulated'. He is unspeakable

[2] *A Descriptive Catalogue.* Number IV.

not by being indefinite but by being too definite for the unavoidable vagueness of language. The words *incorporeal* and *impersonal* are misleading, because they suggest that He lacks some reality which we possess. It would be safer to call Him *trans-corporeal, trans-personal.* Body and personality as we know them are the real negatives—they are what is left of positive being when it is sufficiently diluted to appear in temporal or finite forms. Even our sexuality should be regarded as the transposition into a minor key of that creative joy which in Him is unceasing and irresistible. Grammatically the things we say of Him are 'metaphorical': but in a deeper sense it is our physical and psychic energies that are mere 'metaphors' of the real Life which is God. Divine Sonship is, so to speak, the solid of which biological sonship is merely a diagrammatic representation on the flat.

And here the subject of imagery, which crossed our path in the last chapter, can be seen in a new light. For it is just the recognition of God's positive and concrete reality which the religious imagery preserves. The crudest Old Testament picture of Jahweh thundering and lightning out of dense smoke, making mountains skip like rams, threatening, promising, pleading, even changing His mind, transmits that sense of *living* Deity which evaporates in abstract thought. Even sub-Christian images—even a Hindoo idol with a hundred hands—gets in *something* which mere 'religion' in our own days has left out. We rightly reject it, for by itself it would encourage the most blackguardly of superstitions, the adoration of mere power. Perhaps we may rightly reject much of the Old Testament imagery. But we must be clear why we are doing so: not because the images are too strong but because they are too weak. The ultimate spiritual reality is not vaguer, more inert, more transparent than the images, but more positive, more dynamic, more opaque. Confusion between Spirit and soul (or 'ghost') has here done much harm. Ghosts must be pictured, if we are to picture them at all, as shadowy and tenuous, for ghosts are half-men, one element abstracted from a creature that ought to have flesh. But Spirit, if pictured at all, must be pictured in the very opposite way. Neither God nor even the gods are 'shadowy' in traditional imagination: even the human dead, when glorified in Christ, cease to be 'ghosts' and become 'saints'. The difference of atmosphere which even now surrounds the words 'I saw a ghost' and the words 'I saw a saint'—all the pallor and insubstantiality of the one, all the gold and blue of the other—contains more wisdom than whole libraries of 'religion'. If we must have a mental picture to symbolise Spirit, we should represent it as something *heavier* than matter.

And if we say that we are rejecting the old images in order to do more justice to the moral attributes of God, we must again be careful of what we are really meaning. When we wish to learn of the love and goodness of God by *analogy*—by imagining parallels to them in the realm of human relations—we turn of course to the parables of Christ. But when we try to conceive the reality as it may be in itself, we must beware lest we interpret 'moral attributes' in terms of mere conscientiousness or abstract benevolence. The mistake is easily made because we (correctly) deny that God has passions; and with us a love that is not passionate means a love that is something less. But the reason why God has no passions is that passions imply passivity and intermission. The passion of love is something that happens to us, as 'getting wet' happens to a body: and God is exempt from that 'passion' in the same way that water is exempt from 'getting wet'. He cannot be affected with love, because He *is* love. To imagine that love as something less torrential or less sharp than our own temporary and derivative 'passions' is a most disastrous fantasy.

Again, we may find a violence in some of the traditional imagery which tends to obscure the changelessness of God, the peace, which nearly all who approach Him have reported—the 'still, small voice'. And it is here, I think, that the pre-Christian imagery is least suggestive. Yet even here, there is a danger lest the half conscious picture of some huge thing at rest—a clear, still ocean, a dome of 'white radiance'—should smuggle in ideas of inertia or vacuity. The stillness in which the mystics approach Him is intent and alert—at the opposite pole from sleep or reverie. They are becoming like Him. Silences in the physical world occur in empty places: but the ultimate Peace is silent through very density of life. Saying is swallowed up in being. There is no movement because His action (which is Himself) is timeless. You might, if you wished, call it movement at an infinite speed, which is the same thing as rest, but reached by a different—perhaps a less misleading—way of approach.

Men are reluctant to pass over from the notion of an abstract and negative deity to the living God. I do not wonder. Here lies the deepest tap-root of Pantheism and of the objection to traditional imagery. It was hated not, at bottom, because it pictured Him as man but because it pictured Him as king, or even as warrior. The Pantheist's God does nothing, demands nothing. He is there if you wish for Him, like a book on a shelf. He will not pursue you. There is no danger that at any time heaven and earth should flee away at His glance. If He were the truth, then we could really say that all the Christian images of kingship were a historical accident of which our religion ought to be cleansed. It is with a shock that we discover them to be indispensable. You have had a shock like that before, in connection with smaller matters—when the line pulls at your hand, when something breathes beside you in the darkness. So here; the shock comes at the precise moment when the thrill of *life* is communicated to us along the clue we have been following. It is always shocking to meet life where we thought we were alone. 'Look out!' we cry, 'it's *alive*'. And therefore this is the very point at which so many draw back—I would have done so myself if I could—and proceed no further with Christianity. An 'impersonal God'—well and good. A subjective God of beauty, truth and goodness, inside our own heads—better still. A formless life-force surging through us, a vast power which we can tap—best of all. But God Himself, alive, pulling at the other end of the cord, perhaps approaching at an infinite speed, the hunter, king, husband—that is quite another matter. There comes a moment when the children who have been playing at burglars hush suddenly: was that a *real* footstep in the hall? There comes a moment when people who have been dabbling in religion ('Man's search for God!') suddenly draw back. Supposing we really found Him? We never meant it to come to *that!* Worse still, supposing He had found us?

So it is a sort of Rubicon. One goes across; or not. But if one does, there is no manner of security against miracles. One may be in for *anything*.

12

THE PROPRIETY OF MIRACLES

The Principle at the same moment that it explains the Rules supersedes them.

SEELEY, *Ecce Homo*, chap. xvi.

If the ultimate Fact is not an abstraction but the living God, opaque by the very fullness of His blinding actuality, then He might do things. He might work miracles. But would He? Many people of sincere piety feel that He would not. They think it unworthy of Him. It is petty and capricious tyrants who break their own laws: good and wise kings obey them. Only an incompetent workman will produce work which needs to be interfered with. And people who think in this way are not satisfied by the assurance given them in Chapter VIII that miracles do not, in fact, break the laws of Nature. That may be undeniable. But it will still be felt (and justly) that miracles interrupt the orderly march of events, the steady development of Nature according to her own inherent genius or character. That regular march seems to such critics as I have in mind more impressive than any miracle. Looking up (like Lucifer in Meredith's sonnet) at the night sky, they feel it almost impious to suppose that God should sometimes unsay what He has once said with such magnificence. This feeling springs from deep and noble sources in the mind and must always be treated with respect. Yet it is, I believe, founded on an error.

When schoolboys begin to be taught to make Latin verses at school they are very properly forbidden to have what is technically called 'a spondee in the fifth foot'. It is a good rule for boys because the normal hexameter does not have a spondee there: if boys were allowed to use this abnormal form they would be constantly doing it for convenience and might never get the typical music of the hexameter into their heads at all. But when the boys come to read Virgil they find that Virgil does the very thing they have been forbidden to do—not very often, but not so very rarely either. In the same way, young people who have just learned how to write English rhyming verse, may be shocked at finding 'bad' rhymes (i.e. half-rhymes) in the great poets. Even in carpentry or car-driving or surgery there are, I expect, 'licenses'—abnormal ways of doing things—which the master will use himself both safely and judiciously but which he would think it unwise to teach his pupils.

Now one often finds that the beginner, who has just mastered the strict formal rules, is over-punctilious and pedantic about them. And the mere critic, who is never going to begin himself, may be more pedantic still. The classical critics were shocked at the 'irregularity' or 'licenses' of Shakespeare. A stupid schoolboy might think that the abnormal hexameters in Virgil, or the half-rhymes in English poets, were due to incompetence. In reality, of course, every one of them is there for a purpose and breaks the superficial regularity of the metre in obedience to a higher and subtler law: just as the irregularities in *The Winter's Tale* do not impair, but embody and perfect, the inward unity of its spirit.

In other words, there are rules behind the rules, and a unity which is deeper than uniformity. A supreme workman will never break by one note or one syllable or one stroke of the brush the living and inward law of the work he is producing. But he will break without scruple any number of those superficial regularities and orthodoxies which little, unimaginative critics mistake for its laws. The extent to which one can distinguish a just 'license' from a mere botch or failure of unity depends on the extent to which one has grasped the real and inward significance of the work as a whole. If we had grasped as a whole the innermost spirit of that 'work which God worketh from the beginning to the end', and of which Nature is only a part and perhaps a small part, we should be in a position to decide whether miraculous interruptions of Nature's history were mere improprieties unworthy of the Great Workman or expressions of the truest and deepest unity in His total work. In fact, of course, we are in no such position. The gap between God's mind and ours must, on any view, be incalculably greater than the gap between Shakespeare's mind and that of the most peddling critics of the old French school.

For who can suppose that God's external act, seen from within, would be that same complexity of mathematical relations which Nature, scientifically studied, reveals? It is like thinking that a poet builds up his line out of those metrical feet into which we can analyse it, or that living speech takes grammar as its starting point. But the best illustration of all is Bergson's. Let us suppose a race of people whose peculiar mental limitation compels them to regard a painting as something made up of little coloured dots which have been put together like a mosaic. Studying the brushwork of a great painting, through their magnifying glasses, they discover more and more complicated relations between the dots, and sort these relations out, with great toil, into certain regularities. Their labour will not be in vain. These regularities will in fact 'work'; they will cover most of the facts. But if they go on to conclude that any departure from them would be unworthy of the painter, and an arbitrary breaking of his own rules, they will be far astray. For the regularities they have observed never were the rule the painter was following. What they painfully reconstruct from a million dots, arranged in an agonising complexity, he really produced with a single lightning-quick turn of the wrist, his eye meanwhile taking in the canvas as a whole and his mind obeying laws of composition which the observers, counting their dots, have not yet come within sight of, and perhaps never will. I do not say that the normalities of Nature are unreal. The living fountain of divine energy, solidified for purposes of this spatio-temporal Nature into bodies moving in space and time, and thence, by our abstract thought, turned into mathematical formulae, does in fact for us, commonly fall into such and such patterns. In finding out those patterns we are therefore gaining real, and often useful, knowledge. But to think that a disturbance of them would constitute a breach of the living rule and organic unity whereby God, from His own point of view, works, is a mistake. If miracles do occur then we may be sure that *not* to have wrought them would be the real inconsistency.

How a miracle can be no inconsistency, but the highest consistency, will be clear to those who have read Miss

Dorothy Sayers' indispensable book, *The Mind of the Maker.* Miss Sayers' thesis is based on the analogy between God's relation to the world, on the one hand, and an author's relation to his book on the other. If you are writing a story, miracles or abnormal events may be bad art, or they may not. If, for example, you are writing an ordinary realistic novel and have got your characters into a hopeless muddle, it would be quite intolerable if you suddenly cut the knot and secured a happy ending by having a fortune left to the hero from an unexpected quarter. On the other hand there is nothing against taking as your subject from the outset the adventures of a man who inherits an unexpected fortune. The unusual event is perfectly permissible if it is what you are really writing *about:* it is an artistic crime if you simply drag it in by the heels to get yourself out of a hole. The ghost story is a legitimate form of art; but you must not bring a ghost into an ordinary novel to get over a difficulty in the plot. Now there is no doubt that a great deal of the modern objection to miracles is based on the suspicion that they are marvels of the wrong sort; that a story of a certain kind (Nature) is arbitrarily interfered with, to get the characters out of a difficulty, by events that do not really belong to that kind of story. Some people probably think of the Resurrection as a desperate last moment expedient to save the Hero from a situation which had got out of the Author's control.

The reader may set his mind at rest. If I thought miracles were like that, I should not believe in them. If they have occurred, they have occurred because they are the very thing this universal story is about. They are not exceptions (however rarely they occur) not irrelevancies. They are precisely those chapters in this great story on which the plot turns. Death and Resurrection are what the story is about; and had we but eyes to see it, this has been hinted on every page, met us, in some disguise, at every turn, and even been muttered in conversations between such minor characters (if they are minor characters) as the vegetables. If you have hitherto disbelieved in miracles, it is worth pausing a moment to consider whether this is not chiefly because you thought you had discovered what the story was really about?—that atoms, and time and space and economics and politics were the main plot? And is it certain you were right? It is easy to make mistakes in such matters. A friend of mine wrote a play in which the main idea was that the hero had a pathological horror of trees and a mania for cutting them down. But naturally other things came in as well; there was some sort of love story mixed up with it. And the trees killed the man in the end. When my friend had written it, he sent it an older man to criticise. It came back with the comment, 'Not bad. But I'd cut out those bits of *padding* about the trees'. To be sure, God might be expected to make a better story than my friend. But it is a very *long* story, with a complicated plot; and we are not, perhaps, very attentive readers.

13

ON PROBABILITY

> Probability is founded on the presumption of a resemblance between those objects of which we have had experience and those of which we have had none; and therefore it is impossible that this presumption can arise from probability.
>
> HUME, *Treatise of Human Nature*, I, iii,vi.

The argument up to date shows that miracles are possible and that there is nothing antecedently ridiculous in the stories which say that God has sometimes performed them. This does not mean, of course, that we are committed to believing all stories of miracles. Most stories about miraculous events are probably false: if it comes to that, most stories about natural events are false. Lies, exaggerations, misunderstandings and hearsay make up perhaps more than half of all that is said and written in the world. We must therefore find a criterion whereby to judge any particular story of the miraculous.

In one sense, of course, our criterion is plain. Those stories are to be accepted for which the historical evidence is sufficiently good. But then, as we saw at the outset, the answer to the question, 'How much evidence should we require for this story?' depends on our answer to the question, 'How far is this story intrinsically probable?' We must therefore find a criterion of probability.

The ordinary procedure of the modern historian, even if he admits the possibility of miracle, is to admit no particular instance of it until every possibility of 'natural' explanation has been tried and failed. That is, he will accept the most improbable 'natural' explanations rather than say that a miracle occurred. Collective hallucination, hypnotism of unconsenting spectators, widespread instantaneous conspiracy in lying by persons not otherwise known to be liars and not likely to gain by the lie—all these are known to be very improbable events: so improbable that, except for the special purpose of excluding a miracle, they are never suggested. But they are preferred to the admission of a miracle.

Such a procedure is, from the purely historical point of view, sheer midsummer madness *unless* we start by knowing that any Miracle whatever is more improbable than the most improbable natural event. Do we know this?

We must distinguish the different kinds of improbability. Since miracles are, by definition, rarer than other events, it is obviously improbable beforehand that one will occur at any given place and time. In that sense every miracle is improbable. But that sort of improbability does not make the story that a miracle *has* happened incredible; for in the same sense all events whatever were once improbable. It is immensely improbable beforehand that a pebble dropped from the stratosphere over London will hit any given spot or that any one particular person will win a large lottery. But the report that the pebble has landed outside such and such a shop or that Mr So-and-So has won the lottery is not at all incredible. When you consider the immense number of meetings and fertile unions between ancestors which were necessary in order that you should be born, you perceive that it was once immensely improbable that such a person as you should come to exist: but once you are here, the report of your existence is not in the least incredible. With probability of this kind—antecedent probability of chances—we are not here concerned. Our business is with historical probability.

Ever since Hume's famous *Essay* it has been believed that historical statements about miracles are the most intrinsically improbable of all historical statements. According to Hume, probability rests on what may be called the majority vote of our past experiences. The more often a thing has been known to happen, the more probable it is that it should happen again; and the less often the less probable. Now the regularity of Nature's course, says Hume, is supported by something better than the majority vote of past experiences: it is supported by their unanimous vote, or, as Hume says, by 'firm and unalterable experience'. There is, in fact, 'uniform experience' against Miracle; otherwise, says Hume, it would not be a Miracle. A miracle is therefore the most improbable of all events. It is always more probable that the witnesses were lying or mistaken than that a miracle occurred.

Now of course we must agree with Hume that if there is absolutely 'uniform experience' against miracles, if in other words they have never happened, why then they never have. Unfortunately we know the experience against them to be uniform only if we know that all the reports of them are false. And we can know all the reports to be false only if we know already that miracles have never occurred. In fact, we are arguing in a circle.

There is also an objection to Hume which leads us deeper into our problem. The whole idea of Probability (as Hume understands it) depends on the principle of the Uniformity of Nature. Unless Nature always goes on in the same way, the fact that a thing had happened ten million times would not make it a whit more probable that it would happen again. And how do we know the Uniformity of Nature? A moment's thought shows that we do not know it by experience. We observe many regularities in Nature. But of course all the observations that men have made or will make while the race lasts cover only a minute fraction of the events that actually go on. Our observations would therefore be of no use unless we felt sure that Nature when we are not watching her behaves in the same way as when we are: in other words, unless we believed in the Uniformity of Nature. Experience therefore cannot prove uniformity, because uniformity has to be assumed before experience proves anything. And mere length of experience does not help matters. It is no good saying, 'Each fresh experience confirms our belief in uniformity and therefore we reasonably expect that it will always be confirmed'; for that argument works only on the assumption that the future will resemble the past—which is simply the assumption of Uniformity under a new name. Can we say that Uniformity is at any rate very probable? Unfortunately not. We have just seen that all probabilities depend on *it*. Unless Nature is uniform, nothing is either probable or improbable. And clearly the assumption which you have to make before there is any such thing as probability cannot itself be probable.

The odd thing is that no man knew this better than Hume. His *Essay on Miracles* is quite inconsistent with the more radical, and honourable, scepticism of his main work.

The question, 'Do miracles occur?' and the question, 'Is the course of Nature absolutely uniform?' are the same question asked in two different ways. Hume, by sleight of hand, treats them as two different questions. He first answers 'Yes,' to the question whether Nature is absolutely uniform: and then uses this 'Yes' as a ground for answering, 'No,' to the question, 'Do miracles occur?' The single real question which he set out to answer is never discussed at all. He gets the answer to one form of the question by assuming the answer to another form of the same question.

Probabilities of the kind that Hume is concerned with hold inside the framework of an assumed Uniformity of Nature. When the question of miracles is raised we are asking about the validity or perfection of the frame itself. No study of probabilities inside a given frame can ever tell us how probable it is that the frame itself can be violated. Granted a school timetable with French on Tuesday morning at ten o'clock, it is really probable that Jones, who always skimps his French preparation, will be in trouble next Tuesday, and that he was in trouble on any previous Tuesday. But what does this tell us about the probability of the timetable's being altered? To find that out you must eavesdrop in the masters' common-room. It is no use studying the timetable.

If we stick to Hume's method, far from getting what he hoped (namely, the conclusion that all miracles are infinitely improbable) we get a complete deadlock. The only kind of probability he allows holds exclusively within the frame of uniformity. When uniformity is itself in question (and it is in question the moment we ask whether miracles occur) this kind of probability is suspended. And Hume knows no other. By his method, therefore, we cannot say that uniformity is either probable or improbable; and equally we cannot say that miracles are either probable or improbable. We have impounded *both* uniformity *and* miracles in a sort of limbo where probability and improbability can never come. This result is equally disastrous for the scientist and the theologian; but along Hume's lines there is nothing whatever to be done about it.

Our only hope, then, will be to cast about for some quite different kind of Probability. Let us for the moment cease to ask what right we have to believe in the Uniformity of Nature and ask why in fact men do believe in it. I think the belief has three causes, two of which are irrational. In the first place we are creatures of habit. We expect new situations to resemble old ones. It is a tendency which we share with animals; one can see it working, often to very comic results, in our dogs and cats. In the second place, when we plan our actions, we have to leave out of account the theoretical possibility that Nature might not behave as usual tomorrow, because we can do nothing about it. It is not worth bothering about because no action can be taken to meet it. And what we habitually put out of our minds we soon forget. The picture of uniformity thus comes to dominate our minds without rival and we believe it. Both these causes are irrational and would be just as effective in building up a false belief as in building up a true one.

But I am convinced that there is a third cause. 'In science,' said the late Sir Arthur Eddington, 'we sometimes have convictions which we cherish but cannot justify; we are influenced by some innate sense of the fitness of things'. This may sound a perilously subjective and aesthetic criterion; but can one doubt that it is a principal source of our belief in Uniformity? A universe in which unprecedented and unpredictable events were at every moment flung into Nature would not merely be inconvenient to us: it would be profoundly repugnant. We will not accept such a universe on any terms whatever. It is utterly detestable to us. It shocks our 'sense of the fitness of things'. In advance of experience, in the teeth of many experiences, we are already enlisted on the side of uniformity. For of course science actually proceeds by concentrating not on the regularities of Nature but on her apparent irregularities. It is the apparent irregularity that prompts each new hypothesis. It does so because we refuse to acquiesce in irregularities: we never rest till we have formed and verified a hypothesis which enables us to say that they were not really irregularities at all. Nature as it comes to us looks at first like a mass of irregularities. The stove which lit all right yesterday won't light today; the water which was wholesome last year is poisonous this year. The whole mass of seemingly irregular experience could never have been turned into scientific knowledge at all unless from the very start we had brought to it a faith in uniformity which almost no number of disappointments can shake.

This faith—the preference—is it a thing we can trust? Or is it only the way our minds happen to work? It is useless to say that it has hitherto always been confirmed by the event. That is no good unless you (at least silently) add, 'And therefore always will be': and you cannot add that unless you know already that our faith in uniformity is well grounded. And that is just what we are now asking. Does this sense of fitness of ours correspond to anything in external reality?

The answer depends on the Metaphysic one holds. If all that exists is Nature, the great mindless interlocking event, if our own deepest convictions are merely the by-products of an irrational process, then clearly there is not the slightest ground for supposing that our sense of fitness and our consequent faith in uniformity tell us anything about a reality external to ourselves. Our convictions are simply a fact *about us*—like the colour of our hair. If Naturalism is true we have no reason to trust our conviction that Nature is uniform. It can be trusted only if quite a different Metaphysic is true. If the deepest thing in reality, the Fact which is the source of all other facthood, is a thing in some degree like ourselves—if it is a Rational Spirit and we derive our rational spirituality from It—then indeed our conviction can be trusted. Our repugnance to disorder is derived from Nature's Creator and ours. The disorderly world which we cannot endure to believe in is the disorderly world He would not have endured to create. Our conviction that the timetable will not be perpetually or meaninglessly altered is sound because we have (in a sense) eavesdropped in the Masters' common-room.

The sciences logically require a metaphysic of this sort. Our greatest natural philosopher thinks it is also the metaphysic out of which they originally grew. Professor Whitehead points out[1] that centuries of belief in a God who combined 'the personal energy of Jehovah' with 'the ratio-

[1] *Science and the Modern World*, Chapter II.

nality of a Greek philosopher' first produced that firm expectation of systematic order which rendered possible the birth of modern science. Men became scientific because they expected Law in Nature, and they expected Law in Nature because they believed in a Legislator. In most modern scientists this belief has died: it will be interesting to see how long their confidence in uniformity survives it. Two significant developments have already appeared—the hypothesis of a lawless sub-nature, and the surrender of the claim that science is true. We may be living nearer than we suppose to the end of the Scientific Age.

But if we admit God, must we admit Miracle? Indeed, indeed, you have no security against it. That is the bargain. Theology says to you in effect, 'Admit God and with Him the risk of a few miracles, and I in return will ratify your faith in uniformity as regards the overwhelming majority of events'. The philosophy which forbids you to make uniformity absolute is also the philosophy which offers you solid grounds for believing it to be general, to be *almost* absolute. The Being who threatens Nature's claim to omnipotence confirms her in her lawful occasions. Give us this ha'porth of tar and we will save the ship. The alternative is really much worse. Try to make Nature absolute and you find that her uniformity is not even probable. By claiming too much, you get nothing. You get the deadlock, as in Hume. Theology offers you a working arrangement, which leaves the scientist free to continue his experiments and the Christian to continue his prayers.

We have also, I suggest, found what we were looking for—a criterion whereby to judge the intrinsic probability of an alleged miracle. We must judge it by our 'innate sense of the fitness of things', that same sense of fitness which led us to anticipate that the universe would be orderly. I do not mean, of course, that we are to use this sense in deciding whether miracles in general are possible: we know that they are on philosophical grounds. Nor do I mean that a sense of fitness will do instead of close inquiry into the historical evidence. As I have repeatedly pointed out, the historical evidence cannot be estimated unless we have first estimated the intrinsic probability of the recorded event. It is in making that estimate as regards each story of the miraculous that our sense of fitness comes into play.

If in giving such weight to the sense of fitness I were doing anything new, I should feel rather nervous. In reality I am merely giving formal acknowledgement to a principle which is always used. Whatever men may *say*, no one really thinks that the Christian doctrine of the Resurrection is exactly on the same level with some pious tittle-tattle about how Mother Egarée Louise miraculously found her second best thimble by the aid of St Anthony. The religious and the irreligious are really quite agreed on the point. The whoop of delight with which the sceptic would unearth the story of the thimble, and the 'rosy pudency' with which the Christian would keep it in the background, both tell the same tale. Even those who think all stories of miracles absurd think some very much more absurd than others: even those who believe them all (if anyone does) think that some require a specially robust faith. The criterion which both parties are actually using is that of fitness. More than half the disbelief in miracles that exists is based on a sense of their *unfitness:* a conviction (due, as I have argued, to false philosophy) that they are unsuitable to the dignity of God or Nature or else to the indignity and insignificance of man.

In the three following chapters I will try to present the central miracles of the Christian Faith in such a way as to exhibit their 'fitness'. I shall not, however, proceed by formally setting out the conditions which 'fitness' in the abstract ought to satisfy and then dovetailing the Miracles into that scheme. Our 'sense of fitness' is too delicate and elusive a thing to submit to such treatment. If I succeed, the fitness—and if I fail, the unfitness—of these miracles will of itself become apparent while we study them.

14

THE GRAND MIRACLE

> A light that shone from behind the sun; the sun
> Was not so fierce as to pierce where that light
> could.
>
> CHARLES WILLIAMS

The central miracle asserted by Christians is the Incarnation. They say that God became Man. Every other miracle prepares for this, or exhibits this, or results from this. Just as every natural event is the manifestation at a particular place and moment of Nature's total character, so every particular Christian miracle manifests at a particular place and moment the character and significance of the Incarnation. There is no question in Christianity of arbitrary interferences just scattered about. It relates not a series of disconnected raids on Nature but the various steps of a strategically coherent invasion—an invasion which intends complete conquest and 'occupation'. The fitness, and therefore credibility, of the particular miracles depends on their relation to the Grand Miracle; all discussion of them in isolation from it is futile.

The fitness or credibility of the Grand Miracle itself cannot, obviously, be judged by the same standard. And let us admit at once that it is very difficult to find a standard by which it can be judged. If the thing happened, it was the central event in the history of the Earth—the very thing that the whole story has been about. Since it happened only once, it is by Hume's standards infinitely improbable. But then the whole history of the Earth has also happened only once; is it therefore incredible? Hence the difficulty, which weighs upon Christian and atheist alike, of estimating the probability of the Incarnation. It is like asking whether the existence of Nature herself is intrinsically probable. That is why it is easier to argue, on historical grounds, that the Incarnation actually occurred than to show, on philosophical grounds, the probability of its occurrence. The historical difficulty of giving for the life, sayings and influence of Jesus any explanation that is not harder than the Christian explanation, is very great. The discrepancy between the depth and sanity and (let me add) *shrewdness* of His moral teaching and the rampant megalomania which must lie behind His theological teaching unless He is indeed God, has never been satisfactorily got over.

Hence the non-Christian hypotheses succeed one another with the restless fertility of bewilderment. Today we are asked to regard all the theological elements as later accretions to the story of a 'historical' and merely human Jesus: yesterday we were asked to believe that the whole thing began with vegetation myths and mystery religions and that the pseudo-historical Man was only fadged up at a later date. But this historical inquiry is outside the scope of my book.

Since the Incarnation, if it is a fact, holds this central position, and since we are assuming that we do not yet know it to have happened on historical grounds, we are in a position which may be illustrated by the following analogy. Let us suppose we possess parts of a novel or a symphony. Someone now brings us a newly discovered piece of manuscript and says, 'This is the missing part of the work. This is the chapter on which the whole plot of the novel really turned. This is the main theme of the symphony'. Our business would be to see whether the new passage, if admitted to the central place which the discoverer claimed for it, did actually illuminate all the parts we had already seen and 'pull them together'. Nor should we be likely to go very far wrong. The new passage, if spurious, however attractive it looked at the first glance, would become harder and harder to reconcile with the rest of the work the longer we considered the matter. But if it were genuine then at every fresh hearing of the music or every fresh reading of the book, we should find it settling down, making itself more at home and eliciting significance from all sorts of details in the whole work which we had hitherto neglected. Even though the new central chapter or main theme contained great difficulties in itself, we should still think it genuine provided that it continually removed difficulties elsewhere. Something like this we must do with the doctrine of the Incarnation. Here, instead of a symphony or a novel, we have the whole mass of our knowledge. The credibility will depend on the extent to which the doctrine, if accepted, can illuminate and integrate that whole mass. It is much less important that the doctrine itself should be fully comprehensible. We believe that the sun is in the sky at midday in summer not because we can clearly see the sun (in fact, we cannot) but because we can see everything else.

The first difficulty that occurs to any critic of the doctrine lies in the very centre of it. What can be meant by 'God becoming man'? In what sense is it conceivable that eternal self-existent Spirit, basic Fact-hood, should be so combined with a natural human organism as to make one person? And this would be a fatal stumbling-block if we had not already discovered that in every human being a more than natural activity (the act of reasoning) and therefore presumably a more than natural agent is thus united with a part of Nature: so united that the composite creature calls itself 'I' and 'Me'. I am not, of course, suggesting that what happened when God became Man was simply another instance of this process. In other men a supernatural *creature* thus becomes, in union with the natural creature, one human being. In Jesus, it is held, the Supernatural Creator Himself did so. I do not think anything we do will enable us to imagine the mode of consciousness of the incarnate God. That is where the doctrine is not fully comprehensible. But the difficulty which we felt in the mere idea of the Supernatural descending into the Natural is apparently non-existent, or is at least overcome in the person of every man. If we did not know by experience what it feels like to be a rational animal—how all these natural facts, all this biochemistry and instinctive affection or repulsion and sensuous perception, can become the medium of rational thought and moral will which understand necessary relations and acknowledge modes of behaviour as universally binding, we could not conceive, much less imagine, the thing happening. The discrepancy between a movement of atoms in an astronomer's cortex and his understanding that there must be a still unobserved planet beyond Uranus, is already so immense that the Incarnation of God Himself is, in one sense, scarcely more startling. We cannot conceive how the Divine Spirit dwelled within the created and human spirit of Jesus: but neither can we conceive how His human spirit, or that of any man, dwells within his natural organism. What we can understand, if the Christian doctrine is true, is that our own composite existence is not the sheer anomaly it might seem to be, but a faint image of the Divine Incarnation itself—the same theme in a very minor key. We can understand that if God so descends into a human spirit, and human spirit so descends into Nature, and our thoughts into our senses and passions, and if adult minds (but only the best of them) can descend into sympathy with children, and men into sympathy with beasts, then everything hangs together and the total reality, both Natural and Supernatural, in which we are living is more multifariously and subtly harmonious than we had suspected. We catch sight of a new key principle—the power of the Higher, just in so far as it is truly Higher, to come down, the power of the greater to include the less. Thus solid bodies exemplify many truths of plane geometry, but plane figures no truths of solid geometry: many inorganic propositions are true of organisms but no organic propositions are true of minerals; Montaigne became kittenish with his kitten but she never talked philosophy to him.[1] Everywhere the great enters the little—its power to do so is almost the test of its greatness.

In the Christian story God descends to reascend. He comes down; down from the heights of absolute being into time and space, down into humanity; down further still, if embryologists are right, to recapitulate in the womb ancient and pre-human phases of life; down to the very roots and seabed of the Nature He has created. But He goes down to come up again and bring the whole ruined world up with Him. One has the picture of a strong man stooping lower and lower to get himself underneath some great complicated burden. He must stoop in order to lift, he must almost disappear under the load before he incredibly straightens his back and marches off with the whole mass swaying on his shoulders. Or one may think of a diver, first reducing himself to nakedness, then glancing in mid-air, then gone with a splash, vanished, rushing down through green and warm water into black and cold water, down through increasing pressure into the death-like region of ooze and slime and old decay;

---

[1] *Essays*, I, xii, Apology for Raimond de Sebonde.

then up again, back to colour and light, his lungs almost bursting, till suddenly he breaks surface again, holding in his hand the dripping, precious thing that he went down to recover. He and it are both coloured now that they have come up into the light: down below, where it lay colourless in the dark, he lost his colour too.

In this descent and reascent everyone will recognise a familiar pattern: a thing written all over the world. It is the pattern of all vegetable life. It must belittle itself into something hard, small and deathlike, it must fall into the ground: thence the new life reascends. It is the pattern of all animal generation too. There is descent from the full and perfect organisms into the spermatozoon and ovum, and in the dark womb a life at first inferior in kind to that of the species which is being reproduced: then the slow ascent to the perfect embryo, to the living, conscious baby, and finally to the adult. So it is also in our moral and emotional life. The first innocent and spontaneous desires have to submit to the deathlike process of control or total denial: but from that there is a reascent to fully formed character in which the strength of the original material all operates but in a new way. Death and Rebirth—go down to go up—it is a key principle. Through this bottleneck, this belittlement, the highroad nearly always lies.

The doctrine of the Incarnation, if accepted, puts this principle even more emphatically at the centre. The pattern is there in Nature because it was first there in God. All the instances of it which I have mentioned turn out to be but transpositions of the Divine theme into a minor key. I am not now referring simply to the Crucifixion and Resurrection of Christ. The total pattern, of which they are only the turning point, is the real Death and Rebirth: for certainly no seed ever fell from so fair a tree into so dark and cold a soil as would furnish more than a faint analogy to this huge descent and reascension in which God dredged the salt and oozy bottom of Creation.

From this point of view the Christian doctrine makes itself so quickly at home amid the deepest apprehensions of reality which we have from other sources, that doubt may spring up in a new direction. Is it not fitting in too well? So well that it must have come into men's minds from seeing this pattern elsewhere, particularly in the annual death and resurrection of the corn? For there have, of course, been many religions in which that annual drama (so important for the life of the tribe) was almost admittedly the central theme, and the deity—Adonis, Osiris, or another—almost undisguisedly a personification of the corn, a 'corn-king' who died and rose again each year. Is not Christ simply another corn-king?

Now this brings us to the oddest thing about Christianity. In a sense the view which I have just described is actually true. From a certain point of view Christ is 'the same sort of thing' as Adonis or Osiris (always, of course, waiving the fact that they lived nobody knows where or when, while He was executed by a Roman magistrate we know in a year which can be roughly dated). And that is just the puzzle. If Christianity is a religion of that kind why is the analogy of the seed falling into the ground so seldom mentioned (twice only if I mistake not) in the New Testament? Corn-religions are popular and respectable: if that is what the first Christian teachers were putting across, what motive could they have for concealing the fact? The impression they make is that of men who simply don't know how close they are to the corn-religions: men who simply overlook the rich sources of relevant imagery and association which they must have been on the verge of tapping at every moment. If you say they suppressed it because they were Jews, that only raises the puzzle in a new form. Why should the only religion of a 'dying God' which has actually survived and risen to unexampled spiritual heights occur precisely among those people to whom, and to whom almost alone, the whole circle of ideas that belong to the 'dying God' was foreign? I myself, who first seriously read the New Testament when I was, imaginatively and poetically, all agog for the Death and Rebirth pattern and anxious to meet a corn-king, was chilled and puzzled by the almost total absence of such ideas in the Christian documents. One moment particularly stood out. A 'dying God'—the only dying God who might possibly be historical—holds bread, that is, corn, in His hand and says, 'This is my body'. Surely here, even if nowhere else—or surely if not here, at least in the earliest comments on this passage and through all later devotional usage in ever swelling volume—the truth must come out; the connection between this and the annual drama of the crops must be made. But it is not. It is there for me. There is no sign that it was there for the disciples or (humanly speaking) for Christ Himself. It is almost as if He didn't realise what He had said.

The records, in fact, show us a Person who *enacts* the part of the Dying God, but whose thoughts and words remain quite outside the circle of religious ideas to which the Dying God belongs. The very thing which the Nature-religions are all about seems to have really happened once; but it happened in a circle where no trace of Nature-religion was present. It is as if you met the sea-serpent and found that it disbelieved in sea-serpents: as if history recorded a man who had done all the things attributed to Sir Launcelot but who had himself never apparently heard of chivalry.

There is, however, one hypothesis which, if accepted, makes everything easy and coherent. The Christians are not claiming that simply 'God' was incarnate in Jesus. They are claiming that the one true God is He whom the Jews worshipped as Jahweh, and that it is He who has descended. Now the double character of Jahweh is this. On the one hand He is the God of Nature, her glad Creator. It is He who sends rain into the furrows till the valleys stand so thick with corn that they laugh and sing. The trees of the wood rejoice before Him and His voice causes the wild deer to bring forth their young. He is the God of wheat and wine and oil. In that respect He is constantly doing all the things that Nature-Gods do: He is Bacchus, Venus, Ceres all rolled into one. There is no trace in Judaism of the idea found in some pessimistic and Pantheistic religions that Nature is some kind of illusion or disaster, that finite existence is in itself an evil and that the cure lies in the relapse of all things into God. Compared with such anti-natural conceptions Jahweh might almost be mistaken for a Nature-God.

On the other hand, Jahweh is clearly *not* a Nature-God. He does not die and come to life each year as a true

Corn-king should. He may give wine and fertility, but must not be worshipped with Bacchanalian or aphrodisiac rites. He is not the soul of Nature nor of any part of Nature. He inhabits eternity: He dwells in the high and holy place: heaven is His throne, not His vehicle, earth is His footstool, not His vesture. One day He will dismantle both and make a new heaven and earth. He is not to be identified even with the 'divine spark' in man. He is 'God and not man': His thoughts are not our thoughts: all our righteousness is filthy rags. His appearance to Ezekiel is attended with imagery that does not borrow from Nature, but (it is a mystery too seldom noticed[2]) from those machines which men were to make centuries after Ezekiel's death. The prophet saw something suspiciously like a *dynamo*.

Jahweh is neither the soul of Nature nor her enemy. She is neither His body nor a declension and falling away from Him. She is His creature. He is not a nature-God, but the God of Nature—her inventor, maker, owner, and controller. To everyone who reads this book the conception has been familiar from childhood; we therefore easily think it is the most ordinary conception

---

[2] I owe this point to Canon Adam Fox.

in the world. 'If people are going to believe in a God at all,' we ask, 'what other kind would they believe in?' But the answer of history is, 'Almost any other kind'. We mistake our privileges for our instincts: just as one meets ladies who believe their own refined manners to be natural to them. They don't remember being taught.

Now if there is such a God and if He descends to rise again, then we can understand why Christ is at once so like the Corn-King and so silent about him. He is like the Corn-King because the Corn-King is a portrait of Him. The similarity is not in the least unreal or accidental. For the Corn-King is derived (through human imagination) from the facts of Nature, and the facts of Nature from her Creator; the Death and Rebirth pattern is in her because it was first in Him. On the other hand, elements of Nature-religion are strikingly absent from the teaching of Jesus and from the Judaic preparation which led up to it precisely because in them Nature's Original is manifesting Itself. In them you have from the very outset got in behind Nature-religion and behind Nature herself. Where the real God is present the shadows of that God do not appear; that which the shadows resembled does. The Hebrews throughout their history were being constantly headed off from the worship of Nature-gods; not because the Nature-gods were in all respects unlike the God of Nature but because, at best, they were merely like, and it was the destiny of that nation to be turned away from likenesses to the thing itself.

The mention of that nation turns our attention to one of those features in the Christian story which is repulsive to the modern mind. To be quite frank, we do not at all like the idea of a 'chosen people'. Democrats by birth and education, we should prefer to think that all nations and individuals start level in the search for God, or even that all religions are equally true. It must be admitted at once that Christianity makes no concessions to this point of view. It does not tell of a human search for God at all, but of something done by God for, to, and about, Man. And the way in which it is done is selective, undemocratic, to the highest degree. After the knowledge of God had been universally lost or obscured, one man from the whole earth (Abraham) is picked out. He is separated (miserably enough, we may suppose) from his natural surroundings, sent into a strange country, and made the ancestor of a nation who are to carry the knowledge of the true God. Within this nation there is further selection: some die in the desert, some remain behind in Babylon. There is further selection still. The process grows narrower and narrower, sharpens at last into one small bright point like the head of a spear. It is a Jewish girl at her prayers. All humanity (so far as concerns its redemption) has narrowed to that.

Such a process is very unlike what modern feeling demands: but it is startlingly like what Nature habitually does. Selectiveness, and with it (we must allow) enormous wastage, is her method. Out of enormous space a very small portion is occupied by matter at all. Of all the stars, perhaps very few, perhaps only one, have planets. Of the planets in our own system probably only one supports organic life. In the transmission of organic life, countless seeds and spermatozoa are emitted: some few are selected for the distinction of fertility. Among the species only one is rational. Within that species only a few attain excellence of beauty, strength or intelligence.

At this point we come perilously near the argument of Butler's famous *Analogy*. I say 'perilously' because the argument of that book very nearly admits parodying in the form 'You say that the behaviour attributed to the Christian God is both wicked and foolish: but it is no less likely to be true on that account for I can show that Nature (which He created) behaves just as badly.' To which the atheist will answer—and the nearer he is to Christ in his heart, the more certainly he will do so—'If there is a God like that I despise and defy Him.' But I am not saying that Nature, as we now know her, is good; that is a point we must return to in a moment. Nor am I saying that a God whose actions were no better than Nature's would be a proper object of worship for any honest man. The point is a little finer than that. This selective or undemocratic quality in Nature, at least in so far as it affects human life, is neither good nor evil. According as spirit exploits or fails to exploit this Natural situation, it gives rise to one or the other. It permits, on the one hand, ruthless competition, arrogance, and envy: it permits on the other, modesty and (one of our greatest pleasures) admiration. A world in which I was *really* (and not merely by a useful legal fiction) 'as good as everyone else', in which I never looked up to anyone wiser or cleverer or braver or more learned than I, would be insufferable. The very 'fans' of the cinema stars and the famous footballers know better than to desire that! What the Christian story does is not to instate on the Divine level a cruelty and wastefulness which have already disgusted us on the Natural, but to show us in God's act, working neither cruelly nor wastefully, the same principle which is in Nature also, though down there it works sometimes in one way and sometimes in the other. It illuminates the Natural scene by suggesting that a principle which at first looked meaningless may yet be derived from a principle which is good and fair, may indeed be a depraved and blurred copy of it—the pathological form which it would take in a *spoiled* Nature.

For when we look into the Selectiveness which the Christians attribute to God we find in it none of that 'favouritism' which we were afraid of. The 'chosen' people are chosen not for their own sake (certainly not for their own honour or pleasure) but for the sake of the unchosen. Abraham is told that 'in his seed' (the chosen nation) 'all nations shall be blest'. That nation has been chosen to bear a heavy burden. Their sufferings are great: but, as Isaiah recognised, their sufferings heal others. On the finally selected Woman falls the utmost depth of maternal anguish. Her Son, the incarnate God, is a 'man of sorrows'; the one Man into whom Deity descended, the one Man who can be lawfully adored, is pre-eminent for suffering.

But, you will ask, does this much mend matters? Is not this still injustice, though now the other way round? Where, at the first glance, we accused God of undue favour to His 'chosen', we are now tempted to accuse Him of undue disfavour. (The attempt to keep up both charges at the same time had better be dropped.) And certainly we have here come to a principle very deep-rooted in Christianity: what may be called the principle of *Vicariousness*. The Sinless Man suffers for the sinful, and, in their degree, all good men for all bad men. And this Vicariousness—no less than Death and Rebirth or Selectiveness—is also a characteristic of Nature. Self-sufficiency, living on one's own resources, is a thing impossible in her realm. Everything is indebted to everything else, sacrificed to everything else, dependent on everything else. And here too we must recognise that the principle is in itself neither good nor bad. The cat lives on the mouse in a way I think bad: the bees and the flowers live on one another in a more pleasing manner. The parasite lives on its 'host': but so also the unborn child on its mother. In social life without Vicariousness there would be no exploitation or oppression; but also no kindness or gratitude. It is a fountain both of love and hatred, both of misery and happiness. When we have understood this we shall no longer think that the depraved examples of Vicariousness in Nature forbid us to suppose that the principle itself is of divine origin.

At this point it may be well to take a backward glance and notice how the doctrine of Incarnation is already acting on the rest of our knowledge. We have already brought it into contact with four other principles: the composite nature of man, the pattern of descent and reascension, Selectiveness, and Vicariousness. The first may be called a fact about the frontier between Nature and Supernature; the other three are characteristics of Nature herself. Now most religions, when brought face to face with the facts of Nature either simply reaffirm them, give them (just as they stand) a transcendent prestige, or else simply negate them, promise us release from such facts and from Nature altogether. The Nature-Religions take the first line. They sanctify our agricultural concerns and indeed our whole biological life. We get really drunk in the worship of Dionysus and lie with real women in the temple of the fertility goddess. In Life-force worship, which is the modern and western type of Nature-religion, we take over the existing trend towards 'development' of increasing complexity in organic, social, and industrial life, and make it a god. The anti-Natural or pessimistic religions, which are more civilised and sensitive, such as Buddhism or higher Hinduism, tell us that Nature is evil and illusory, that there is an escape from this incessant change, this furnace of striving and desire. Neither the one nor the other sets the facts of Nature in a new light. The Nature-religions merely reinforce that view of Nature which we spontaneously adopt in our moments of rude health and cheerful brutality; the anti-natural religions do the same for the view we take in moments of compassion, fastidiousness, or lassitude. The Christian doctrine does neither of these things. If any man approaches it with the idea that because Jahweh is the God of fertility our lasciviousness is going to be authorised or that the Selectiveness and Vicariousness of God's method will excuse us for imitating (as 'Heroes', 'Supermen' or social parasites) the lower Selectiveness and Vicariousness of Nature, he will be stunned and repelled by the inflexible Christian demand for chastity, humility, mercy and justice. On the other hand if we come to it regarding the death which precedes every rebirth, or the fact of inequality, or our dependence on others and their dependence on us, as the mere odious necessities of an evil cosmos, and hoping to be delivered into transparent and 'enlightened' spirituality where all these things just vanish, we shall be equally disappointed. We shall be told that, in one sense, and despite enormous differences, it is 'the same all the way up'; that hierarchical inequality, the need for self surrender, the willing sacrifice of self to others, and the thankful and loving (but unashamed) acceptance of others' sacrifice to us, hold sway in the realm beyond Nature. It is indeed only love that makes the difference: all those very same principles which are evil in the world of selfishness and necessity are good in the world of love and understanding. Thus, as we accept this doctrine of the higher world we make new discoveries about the lower world. It is from that hill that we first really understand the landscape of this valley. Here, at last, we find (as we do not find either in the Nature-religions or in the religions that deny Nature) a real illumination: Nature is being lit up by a light from beyond Nature. Someone is speaking who knows more about her than can be known from inside her.

Throughout this doctrine it is, of course, implied that Nature is infested with evil. Those great key-principles which exist as modes of goodness in the Divine Life, take on, in her operations, not merely a less perfect form (that we should, on any view, expect) but forms which I have been driven to describe as morbid or depraved. And this depravity could not be totally removed without the drastic remaking of Nature. Complete human virtue could indeed banish from human life all the evils that now arise in it from Vicariousness and Selectiveness and retain only the good: but the wastefulness and painfulness of non-human Nature would remain—and would, of course, continue to infect human life in the form of disease. And the destiny which Christianity promises to man clearly involves a 'redemption' or 'remaking' of Nature which could not stop at Man, or even at this planet. We are told that 'the whole creation' is in travail, and that Man's rebirth will be the signal for hers. This gives rise to several problems, the discussion of which puts the whole doctrine of the Incarnation in a clearer light.

In the first place, we ask how the Nature created by a good God comes to be in this condition? By which question we may mean either how she comes to be imperfect—to leave 'room for improvement' as the schoolmasters say in their reports—or else, how she comes to be positively depraved. If we ask the question in the first sense, the Christian answer (I think) is that God, from the first, created her such as to reach her perfection by a process in time. He made an Earth at first 'without form and void' and brought it by degrees to its perfection. In this, as elsewhere, we see the familiar pattern—descent from God to the formless Earth and reascent from the formless to the finished. In that sense a certain degree of 'evolutionism' or 'developmentalism' is inherent in Christianity. So much for Nature's imperfection; her positive depravity calls for a very different explanation. According to the Christians this is all due to sin: the sin both of men and of powerful, non-human beings, supernatural but created. The unpopularity of this doctrine arises from the widespread Naturalism of our age—the belief that nothing but Nature exists and that if anything else did she is protected from it by a Maginot Line—and will disappear as this error is corrected. To be sure, the morbid inquisitiveness about such beings which led our ancestors to a pseudo-science of Demonology, is to be sternly discouraged: our attitude should be that of the sensible citizen in wartime who believes that there are enemy spies in our midst but disbelieves nearly every particular spy story. We must limit ourselves to the general statement that beings in a different, and higher 'Nature' which is *partially* interlocked with ours have, like men, fallen and have tampered with things inside our frontier. The doctrine, besides proving itself fruitful of good in each man's spiritual life, helps to protect us from shallowly optimistic or pessimistic views of Nature. To call her either 'good' or 'evil' is boys' philosophy. We find ourselves in a world of transporting pleasures, ravishing beauties, and tantalising possibilities, but all constantly being destroyed, all coming to nothing. Nature has all the air of a good thing spoiled.

The sin, both of men and of angels, was rendered possible by the fact that God gave them free will: thus surrendering a portion of His omnipotence (it is again a deathlike or descending movement) because He saw that from a world of free creatures, even though they fell, He could work out (and this is the reascent) a deeper happiness and a fuller splendour than any world of automata would admit.

Another question that arises is this. If the redemption of Man is the beginning of Nature's redemption as a whole, must we then conclude after all that Man is the most important thing in Nature? If I had to answer 'Yes' to this question I should not be embarrassed. Supposing Man to be the only rational animal in the universe, then (as has been shown) his small size and the small size of the globe he inhabits would not make it ridiculous to regard him as the hero of the cosmic drama: Jack after all is the smallest character in *Jack the Giant-Killer.* Nor do I think it in the least improbable that Man is in fact the only rational creature in this spatio-temporal Nature. That is just the sort of lonely pre-eminence—just the disproportion between picture and frame—which all that I know of Nature's 'selectiveness' would lead me to anticipate. But I do not need to assume that it actually exists. Let Man be only one among a myriad of rational species, and let him be the only one that has fallen. Because he has fallen, for him God does the great deed; just as in the parable it is the one lost sheep for whom the shepherd hunts. Let Man's pre-eminence or solitude be one not of superiority but of misery and evil: then, all the more, Man will be the very species into which Mercy will descend. For this prodigal the fatted calf, or, to speak more suitably, the eternal Lamb, is killed. But once the Son of God, drawn hither not by our merits but by our unworthiness, has put on human nature, then our species (whatever it may have been before) does become in one sense the central fact in all Nature: our species, rising after its long descent, will drag all Nature up with it because in our species the Lord of Nature is now included. And it would be all of a piece with what we already know if ninety and nine righteous races inhabiting distant planets that circle distant suns, and needing no redemption on their own account, were remade and glorified by the glory which had descended into our race. For God is not merely mending, not simply restoring a *status quo.* Redeemed humanity is to be something more glorious than unfallen humanity would have been, more glorious than any unfallen race now is (if at this moment the night sky conceals any such). The greater the sin, the greater the mercy: the deeper the death the brighter the rebirth. And this super-added glory will, with true vicariousness, exalt all creatures and those who have never fallen will thus bless Adam's fall.

I write so far on the assumption that the Incarnation was occasioned only by the Fall. Another view has, of course, been sometimes held by Christians. According to it the descent of God into Nature was not in itself occasioned by sin. It would have occurred for Glorification and Perfection even if it had not been required for Redemption. Its attendant circumstances would have been very different: the divine humility would not have been a divine humiliation, the sorrows, the gall and vinegar, the crown of thorns and the cross, would have been absent. If this view is taken, then clearly the Incarnation, wherever and however it occurred, would always have been the beginning of Nature's rebirth. The fact that it has occurred in the human species, summoned thither by that strong incantation of misery and abjection which Love has made Himself unable to resist, would not deprive it of its universal significance.

This doctrine of a universal redemption spreading outwards from the redemption of Man, mythological as it will seem to modern minds, is in reality far more philosophical than any theory which holds that God, having once entered Nature, should leave her, and leave her substantially unchanged, or that the glorification of one creature could be realised without the glorification of the whole system. God never undoes anything but evil, never does good to undo it again. The union between God and Nature in the Person of Christ admits no divorce. He will not *go out* of Nature again and she must be glorified in all ways which this miraculous union demands. When spring comes it 'leaves no corner of the land untouched'; even a pebble dropped in a pond sends circles to the margin. The question we want to ask about Man's 'central' position in this drama is really on a level with the disciples' question, 'Which of them was the greatest?' It is the sort of question which God does not answer. If from Man's point of view the re-creation of non-human and even inanimate Nature appears a mere by-product of his own redemption, then equally from some remote, non-human point of view Man's redemption may seem merely the preliminary to this more widely diffused springtime, and the very permission of Man's fall may be supposed to have had that larger end in view. Both attitudes will be right if they will consent to drop the words *mere* and *merely.* Where a God who is totally purposive and totally foreseeing acts upon a Nature which is totally interlocked, there can be no accidents or loose ends, nothing whatever of which we can safely use the word *merely.* Nothing is 'merely a by-product' of anything else. All results are intended from the first. What is subservient from one point of view is the main purpose from another. No thing or event is first or highest in a sense which forbids it to be also last and lowest. The partner who bows to Man in one movement of the dance receives Man's reverences in another. To be high or central means to abdicate continually: to be low means to be raised: all good masters are servants: God washes the feet of men. The concepts we usually bring to the consideration of such matters are miserably political and prosaic. We think of flat repetitive equality and arbitrary privilege as the only two alternatives—thus missing all the overtones, the counterpoint, the vibrant sensitiveness, the inter-inanimations of reality.

For this reason I do not think it at all likely that there have been (as Alice Meynell suggested in an interesting poem) many Incarnations to redeem many different kinds of creature. One's sense of *style*—of the divine idiom—rejects it. The suggestion of mass-production and of waiting queues comes from a level of thought which is here hopelessly inadequate. If other natural creatures than Man have sinned we must believe that they are redeemed: but God's Incarnation as Man will be one unique act in the drama of total redemption and other species will have witnessed wholly different acts, each equally unique, equally necessary and differently necessary to the whole process, and each (from a certain point of view) justifiably regarded as 'the great scene' of the play. To those who live in Act II, Act III looks like an epilogue: to those who live in Act III, Act II looks like a prologue. And both are right until they add the fatal word *merely*, or else try to avoid it by the dullard's supposition that both acts are the same.

It ought to be noticed at this stage that the Christian doctrine, if accepted, involves a particular view of Death. There are two attitudes towards Death which the human mind naturally adopts. One is the lofty view, which reached its greatest intensity among the Stoics, that Death 'doesn't matter', that it is 'kind nature's signal for retreat', and that we ought to regard it with indifference. The other is the 'natural' point of view, implicit in nearly all private conversations on the subject, and in much modern thought about the survival of the human species, that Death is the greatest of all evils: Hobbes is perhaps the only philosopher who erected a system on this basis. The first idea simply negates, the second simply affirms, our instinct for self-preservation; neither throws any new light on Nature, and Christianity countenances neither. Its doctrine is subtler. On the one hand Death is the triumph of Satan, the punishment of the Fall, and the last enemy. Christ shed tears at the grave of Lazarus and sweated blood in Gethsemane: the Life of Lives that was in Him detested this penal obscenity not less than we do, but more. On the other hand, only he who loses his life will save it. We are baptised into the *death* of Christ, and it is the remedy for the Fall. Death is, in fact, what some modern people call 'ambivalent'. It is Satan's great weapon and also God's great weapon: it is holy and unholy; our supreme disgrace and our only hope; the thing Christ came to conquer and the means by which He conquered.

To penetrate the whole of this mystery is, of course, far beyond our power. If the pattern of Descent and Reascent is (as looks not unlikely) the very formula of reality, then in the mystery of Death the secret of secrets lies hid. But something must be said in order to put the Grand Miracle in its proper light. We need not discuss Death on the highest level of all: the mystical slaying of the Lamb 'before the foundation of the world' is above our speculations. Nor need we consider Death on the lowest level. The death of organisms which are nothing more than organisms, which have developed no personality, does not concern us. Of it we may truly say, as some spiritually minded people would have us say of human Death, that it 'doesn't matter'. But the startling Christian doctrine of human Death cannot be passed over.

Human Death, according to the Christians, is a result of human sin; Man, as originally created, was immune from it: Man, when redeemed, and recalled to a new life (which will, in some undefined sense, be a bodily life) in the midst of a more organic and more fully obedient Nature, will be immune from it again. This doctrine is of course simply nonsense if a man is nothing but a Natural organism. But if he were, then, as we have seen, all thoughts would be equally nonsensical, for all would have irrational causes. Man must therefore be a composite being—a natural organism tenanted by, or in a state of *symbiosis* with, a supernatural spirit. The Christian doctrine, startling as it must seem to those who have not fully cleared their minds of Naturalism, states that the relations which we now observe between that spirit and that organism, are abnormal or pathological ones. At present spirit can retain its foothold against the incessant counter-attacks of Nature (both physiological and psychological) only by perpetual vigilance, and physiological Nature always defeats it in the end. Sooner or later it becomes unable to resist the disintegrating processes at work in the body and death ensues. A little later the Natural organism (for it does not long enjoy its triumph) is similarly conquered by merely physical Nature and returns to the inorganic. But, on the Christian view, this was not always so. The spirit was once not a garrison, maintaining its post with difficulty in a hostile

Nature, but was fully 'at home' with its organism, like a king in his own country or a rider on his own horse—or better still, as the human part of a Centaur was 'at home' with the equine part. Where spirit's power over the organism was complete and unresisted, death would never occur. No doubt, spirit's permanent triumph over natural forces which, if left to themselves, would kill the organism, would involve a continued miracle: but only the same sort of miracle which occurs every day—for whenever we think rationally we are, by direct spiritual power, forcing certain atoms in our brain and certain psychological tendencies in our natural soul to do what they would never have done if left to Nature. The Christian doctrine would be fantastic only if the present frontier-situation between spirit and Nature in each human being were so intelligible and self-explanatory that we just 'saw' it to be the only one that could ever have existed. But is it?

In reality the frontier situation is so odd that nothing but custom could make it seem natural, and nothing but the Christian doctrine can make it fully intelligible. There is certainly a state of war. But not a war of mutual destruction. Nature by dominating spirit wrecks all spiritual activities: spirit by dominating Nature confirms and improves natural activities. The brain does not become less a brain by being used for rational thought. The emotions do not become weak or jaded by being organised in the service of a moral will—indeed they grow richer and stronger as a beard is strengthened by being shaved or a river is deepened by being banked. The body of the reasonable and virtuous man, other things being equal, is a better body than that of the fool or the debauchee, and his sensuous pleasures better simply as sensuous pleasures: for the slaves of the senses, after the first bait, are starved by their masters. Everything happens as if what we saw was not war, but rebellion: that rebellion of the lower against the higher by which the lower destroys both the higher and itself. And if the present situation is one of rebellion, then reason cannot reject but will rather demand the belief that there was a time before the rebellion broke out and may be a time after it has been settled. And if we thus see grounds for believing that the supernatural spirit and the natural organism in Man have quarrelled, we shall immediately find it confirmed from two quite unexpected quarters.

Almost the whole of Christian theology could perhaps be deduced from the two facts (*a*) That men make coarse jokes, and (*b*) That they feel the dead to be uncanny. The coarse joke proclaims that we have here an animal which finds its own animality either objectionable or funny. Unless there had been a quarrel between the spirit and the organism I do not see how this could be: it is the very mark of the two not being 'at home' together. But is very difficult to imagine such a state of affairs as original—to suppose a creature which from the very first was half shocked and half tickled to death at the mere fact of being the creature it is. I do not perceive that dogs see anything funny about being dogs: I suspect that angels see nothing funny about being angels. Our feeling about the dead is equally odd. It is idle to say that we dislike corpses because we are afraid of ghosts. You might say with equal truth that we fear ghosts because we dislike corpses—for the ghost owes much of its horror to the associated ideas of pallor, decay, coffins, shrouds, and worms. In reality we hate the division which makes possible the conception of either corpse or ghost. Because the thing ought not to be divided, each of the halves into which it falls by division is detestable. The explanations which Naturalism gives both of bodily shame and of our feeling about the dead are not satisfactory. It refers us to primitive taboos and superstitions—as if these themselves were not obviously results of the thing to be explained. But once accept the Christian doctrine that man was originally a unity and that the present division is unnatural, and all the phenomena fall into place. It would be fantastic to suggest that the doctrine was devised to explain our enjoyment of a chapter in Rabelais, a good ghost story, or the *Tales* of Edgar Allan Poe. It does so none the less.

I ought, perhaps, to point out that the argument is not in the least affected by the value-judgements we make about ghost stories or coarse humour. You may hold that both are bad. You may hold that both, though they result (like clothes) from the Fall, are (like clothes) the proper way to deal with the Fall once it has occurred: that while perfected and recreated Man will no longer experience that kind of laughter or that kind of shudder, yet here and now not to feel the horror and not to see the joke is to be less than human. But either way the facts bear witness to our present maladjustment.

So much for the sense in which human Death is the result of sin and the triumph of Satan. But it is also the means of redemption from sin, God's medicine for Man and His weapon against Satan. In a general way it is not difficult to understand how the same thing can be a masterstroke on the part of one combatant and also the very means whereby the superior combatant defeats him. Every good general, every good chess-player, takes what is precisely the strong point of his opponent's plan and makes it the pivot of his own plan. Take that castle of mine if you insist. It was not my original intention that you should—indeed, I thought you would have had more sense. But take it by all means. For now I move thus . . . and thus . . . and it is mate in three moves. Something like this must be supposed to have happened about Death. Do not say that such metaphors are too trivial to illustrate so high a matter: the unnoticed mechanical and mineral metaphors which, in this age, will dominate our whole minds (without being recognised as metaphors at all) the moment we relax our vigilance against them, must be incomparably less adequate.

And one can see how it might have happened. The Enemy persuades Man to rebel against God: Man, by doing so, loses power to control that other rebellion which the Enemy now raises in Man's organism (both psychical and physical) against Man's spirit: just as that organism, in its turn, loses power to maintain itself against the rebellion of the inorganic. In that way, Satan produced human Death. But when God created Man he gave him such a constitution that, if the highest part of it rebelled against Himself, it would be bound to lose control over the lower parts: i.e. in the long run to suffer Death. This provision may be regarded equally as a punitive sentence ('In the day ye eat of that fruit ye shall die'), as a mercy, and as a safety device. It is punishment because Death—that Death of which Martha says to Christ 'But . . . Sir . . . it'll *smell*'—is horror and ignominy. ('I am not so much afraid of death as ashamed of it,' said Sir Thomas Browne). It is mercy because by willing and humble surrender to it Man undoes his act of rebellion and makes even this depraved and monstrous mode of Death an instance of that higher and mystical Death which is eternally good and a necessary ingredient in the highest life. 'The readiness is all'—not, of course, the merely heroic readiness but that of humility and self-renunciation. Our enemy, so welcomed, becomes our servant: bodily Death, the monster, becomes blessed spiritual Death to self, if the spirit so wills—or rather if it allows the Spirit of the willingly dying God so to will in it. It is a safety-device because, once Man has fallen, natural immortality would be the one utterly hopeless destiny for him. Aided to the surrender that he must make by no external necessity of Death, free (if you call it freedom) to rivet faster and faster about himself through unending centuries the chains of his own pride and lust and of the nightmare civilisations which these build up in ever-increasing power and complication, he would progress from being merely a fallen man to being a fiend, possibly beyond all modes of redemption. This danger was averted. The sentence that those who ate of the forbidden fruit would be driven away from the Tree of Life was implicit in the composite nature with which Man was created. But to convert this penal death into the means of eternal life—to add to its negative and preventive function a positive and saving function—it was further necessary that death should be *accepted*. Humanity must embrace death freely, submit to it with total humility, drink it to the dregs, and so convert it into that mystical death which is the secret of life. But only a Man who did not need to have been a Man at all unless He had chosen, only one who served in our sad regiment as a volunteer, yet also only one who was perfectly a Man, could perform this perfect dying; and thus (which way you put it is unimportant) either defeat death or redeem it. He tasted death on behalf of all others. He is the representative 'Die-er' of the universe: and for that very reason the Resurrection and the Life. Or conversely, because He truly lives, He truly dies, for that is the very pattern of reality. Because the higher can descend into the lower He who from all eternity has been incessantly plunging Himself in the blessed death of self-surrender to the Father can also most fully descend into the horrible and (for us) involuntary death of the body. Because Vicariousness is the very idiom of the reality He has created, His death can become ours. The whole Miracle, far from denying what we already know of reality, writes the comment which makes that crabbed text plain: or rather, proves itself to be the text on which Nature was only the commentary. In science we have been reading only the notes to a poem; in Christianity we find the poem itself.

With this our sketch of the Grand Miracle may end. Its credibility does not lie in Obviousness. Pessimism, Optimism, Pantheism, Materialism, all have this 'obvious' attraction. Each is confirmed at the first glance by multitudes of facts: later on, each meets insuperable obstacles. The doctrine of the Incarnation works into our minds quite differently. It digs beneath the surface, works through the rest of our knowledge by unexpected channels, harmonises best with our deepest apprehensions and our 'second thoughts', and in union with these undermines our superficial opinions. It has little to say to the man who is still certain that everything is going to the dogs, or that everything is getting better and better, or that everything is God, or that everything is electricity. Its hour comes when these wholesale creeds have begun to fail us. Whether the thing really happened is a historical question. But when you turn to history, you will not demand for it that kind and degree of evidence which you would rightly demand for something intrinsically improbable; only that kind and degree which you demand for something which, if accepted, illuminates and orders all other phenomena, explains both our laughter and our logic, our fear of the dead and our knowledge that it is somehow good to die, and which at one stroke covers what multitudes of separate theories will hardly cover for us if this is rejected.

15

MIRACLES OF THE OLD CREATION

> The Son can do nothing of himself, but what he seeth the Father do.
>
> John 5:19

If we open such books as Grimm's *Fairy Tales* or Ovid's *Metamorphoses* or the Italian epics we find ourselves in a world of miracles so diverse that they can hardly be classified. Beasts turn into men and men into beasts or trees, trees talk, ships become goddesses, and a magic ring can cause tables richly spread with food to appear in solitary places. Some people cannot stand this kind of story, others find it fun. But the least suspicion that it was true would turn the fun into nightmare. If such things really happened they would, I suppose, show that Nature was being invaded. But they would show that she was being invaded by an alien power. The fitness of the Christian miracles, and their difference from these mythological miracles, lies in the fact that they show invasion by a Power which is not alien. They are what might be expected to happen when she is invaded not simply by a god, but by the God of Nature: by a Power which is outside her jurisdiction not as a foreigner but as a sovereign. They proclaim that He who has come is not merely a king, but *the* King, her King and ours.

It is this which, to my mind, puts the Christian miracles in a different class from most other miracles. I do not think that it is the duty of a Christian apologist (as many sceptics suppose) to disprove all stories of the miraculous which fall outside the Christian records, nor of a Christian man to disbelieve them. I am in no way committed to the assertion that God has never worked miracles through and for Pagans or never permitted created supernatural beings to do so. If, as Tacitus, Suetonius, and Dion Cassius relate, Vespasian performed two cures, and if modern doctors tell me that they could not have been performed without miracle, I have no objection. But I claim that the Christian miracles have a much greater intrinsic probability in virtue of their organic connection with one another and with the whole structure of the religion they exhibit. If it can be shown that one particular Roman emperor—and, let us admit, a fairly good emperor as emperors go—once was empowered to do a miracle, we must of course put up with the fact. But it would remain a quite isolated and anomalous fact. Nothing comes of it, nothing leads up to it, it establishes no body of doctrine, explains nothing, is connected with nothing. And this, after all, is an unusually favourable instance of a non-Christian miracle. The immoral, and sometimes almost idiotic interferences attributed to gods in Pagan stories, even if they had a trace of historical evidence, could be accepted only on the condition of our accepting a wholly meaningless universe. What raises infinite difficulties and solves none will be believed by a rational man only under absolute compulsion. Sometimes the credibility of the miracles is in an inverse ratio to the credibility of the religion. Thus miracles are (in late documents, I believe) recorded of the Buddha. But what could be more absurd than that he who came to teach us that Nature is an illusion from which we must escape should occupy himself in producing effects on the Natural level—that he who comes to wake us from a nightmare should *add* to the nightmare? The more we respect his teaching the less we could accept his miracles. But in Christianity, the more we understand what God it is who is said to be present and the purpose for which He is said to have appeared, the more credible the miracles become. That is why we seldom find the Christian miracles denied except by those who have abandoned some part of the Christian doctrine. The mind which asks for a non-miraculous

Christianity is a mind in process of relapsing from Christianity into mere 'religion'.[1]

The miracles of Christ can be classified in two ways. The first system yields the classes (1) Miracles of Fertility (2)

[1] A consideration of the Old Testament miracles is beyond the scope of this book and would require many kinds of knowledge which I do not possess. My present view—which is tentative and liable to any amount of correction—would be that just as, on the factual side, a long preparation culminates in God's becoming incarnate as Man, so, on the documentary side, the truth first appears in *mythical* form and then by a long process of condensing or focusing finally becomes incarnate as History. This involves the belief that Myth in general is not merely misunderstood history (as Euhemerus thought) nor diabolical illusion (as some of the Fathers thought) nor priestly lying (as the philosophers of the Enlightenment thought) but, at its best, a real though unfocused gleam of divine truth falling on human imagination. The Hebrews, like other people, had mythology: but as they were the chosen people so their mythology was the chosen mythology—the mythology chosen by God to be the vehicle of the earliest sacred truths, the first step in that process which ends in the New Testament where truth has become completely historical. Whether we can ever say with certainty where, in this process of crystallisation, any particular Old Testament story fails, is another matter. I take it that the Memoirs of David's court come at one end of the scale and are scarcely less historical than St Mark or Acts; and that the Book of Jonah is at the opposite end. It should be noted that on this view (*a*) Just as God, in becoming Man, is 'emptied' of His glory, so the truth, when it comes down from the 'heaven' of myth to the 'earth' of history, undergoes a certain humiliation. Hence the New Testament is, and ought to be, more prosaic, in some ways less *splendid*, than the Old; just as the Old Testament is and ought to be less rich in many kinds of imaginative beauty than the Pagan mythologies. (*b*) Just as God is none the less God by being Man, so the Myth remains Myth even when it becomes Fact. The story of Christ demands from us, and repays, not only a religious and historical but also an imaginative response. It is directed to the child, the poet, and the savage in us as well as to the conscience and to the intellect. One of its functions is to break down dividing walls.

Miracles of Healing (3) Miracles of Destruction (4) Miracles of Dominion over the Inorganic (5) Miracles of Reversal (6) Miracles of Perfecting or Glorification. The second system, which cuts across the first, yields two classes only: they are (1) Miracles of the Old Creation, and (2) Miracles of the New Creation.

I contend that in all these miracles alike the incarnate God does suddenly and locally something that God has done or will do in general. Each miracle writes for us in small letters something that God has already written, or will write, in letters almost too large to be noticed, across the whole canvas of Nature. They focus at a particular point either God's actual, or His future, operations on the universe. When they reproduce operations we have already seen on the large scale they are miracles of the Old Creation: when they focus those which are still to come they are miracles of the New. Not one of them is isolated or anomalous: each carries the signature of the God whom we know through conscience and from Nature. Their authenticity is attested by the *style.*

Before going any further I should say that I do not propose to raise the question, which has before now been asked, whether Christ was able to do these things only because He was God or also because He was perfect man; for it is a possible view that if Man had never fallen all men would have been able to do the like. It is one of the glories of Christianity that we can say of this question. 'It doesn't matter.' Whatever may have been the powers of unfallen man, it appears that those of redeemed Man will be almost unlimited.[2] Christ, reascending from His great dive, is bringing up Human Nature with Him. Where He goes, it goes too. It will be made 'like Him'.[3] If in His miracles He is not acting as the Old Man might have done before his Fall, then He is acting as the New Man, every new man, will do after his redemption. When humanity, borne on His shoulders, passes with Him up from the cold dark water into the green warm water and out at last into the sunlight and the air, it also will be bright and coloured.

Another way of expressing the real character of the miracles would be to say that though isolated from other actions, they are not isolated in either of the two ways we are apt to suppose. They are not, on the one hand, isolated from other Divine acts: they do close and small and, as it were, in focus what God at other times does so large that men do not attend to it. Neither are they isolated exactly as

---

[2] Matthew 17:20, 21:21, Mark 11:23, Luke 10:19, John 14:12, 1 Corinthians 3:22, 2 Timothy 2:12.

[3] Philippians 3:21, 1 John 3:1, 2.

we suppose from other human acts: they anticipate powers which all men will have when they also are 'sons' of God and enter into that 'glorious liberty'. Christ's isolation is not that of a prodigy but of a pioneer. He is the first of His kind; He will not be the last.

Let us return to our classification and firstly to Miracles of *Fertility*. The earliest of these was the conversion of water into wine at the wedding feast in Cana. This miracle proclaims that the God of all wine is present. The vine is one of the blessings sent by Jahweh: He is the reality behind the false god Bacchus. Every year, as part of the Natural order, God makes wine. He does so by creating a vegetable organism that can turn water, soil and sunlight into a juice which will, under proper conditions, become wine. Thus, in a certain sense, He constantly turns water into wine, for wine, like all drinks, is but water modified. Once, and in one year only, God, now incarnate, short circuits the process: makes wine in a moment: uses earthenware jars instead of vegetable fibres to hold the water. But uses them to do what He is always doing. The miracle consists in the short cut; but the event to which it leads is the usual one. If the thing happened, then we know that what has come into Nature is no anti-Natural spirit, no God who loves tragedy and tears and fasting *for their own sake* (however He may permit or demand them for special purposes) but the God of Israel who has through all these centuries given us wine to gladden the heart of man.

Other miracles that fall in this class are the two instances of miraculous feeding. They involve the multiplication of a little bread and a little fish into much bread and much fish. Once in the desert Satan had tempted Him to make bread of stones: He refused the suggestion. 'The Son does nothing except what He sees the Father do': perhaps one may without boldness surmise that the direct change from stone to bread appeared to the Son to be not quite in the hereditary style. Little bread into much bread is quite a different matter. Every year God makes a little corn into much corn: the seed is sown and there is an increase. And men say, according to their several fashions, 'It is the laws of Nature,' or 'It is Ceres, it is Adonis, it is the Corn-King'. But the laws of Nature are only a pattern: nothing will come of them unless they can, so to speak, take over the universe as a going concern. And as for Adonis, no man can tell us where he died or when he rose again. Here, at the feeding of the five thousand, is He whom we have ignorantly worshipped: the *real* Corn-King who will die once and rise once at Jerusalem during the term of office of Pontius Pilate.

That same day He also multiplied fish. Look down into every bay and almost every river. This swarming, undulating fecundity shows He is still at work 'thronging the seas with spawn innumerable'. The ancients had a god called Genius; the god of animal and human fertility, the patron of gynaecology, embryology, and the marriage bed—the 'genial' bed as they called it after its god Genius. But Genius is only another mask for the God of Israel, for it was He who at the beginning commanded all species 'to be fruitful and multiply and replenish the earth'. And now, that day, at the feeding of the thousands, incarnate God does the same: does close and small, under His human hands, a workman's hands, what He has always been doing in the seas, the lakes and the little brooks.

With this we stand on the threshold of that miracle which for some reason proves hardest of all for the modern mind to accept. I can understand the man who denies miracles altogether: but what is one to make of people who will believe other miracles and 'draw the line' at the Virgin Birth? Is it that for all their lip service to the laws of Nature there is only one natural process in which they really believe? Or is it that they think they see in this miracle a slur upon sexual intercourse (though they might just as well see in the feeding of the five thousand an insult to bakers) and that sexual intercourse is the one thing still venerated in this unvenerating age? In reality the miracle is no less, and no more, surprising than any others.

Perhaps the best way to approach it is from the remark I saw in one of the most archaic of our anti-god papers. The remark was that Christians believed in a God who had 'committed adultery with the wife of a Jewish carpenter'. The writer was probably merely 'letting off steam' and did not really think that God, in the Christian story, had assumed human form and lain with a mortal woman, as Zeus lay with Alcmena. But if one had to answer this person, one would have to say that if you called the miraculous conception divine adultery you would by driven to find a similar divine adultery in the conception of every child—nay, of every animal too. I am sorry to use expressions which will offend pious ears, but I do not know how else to make my point.

In a normal act of generation the father has no creative function. A microscopic particle of matter from his body, and a microscopic particle from the woman's body, meet. And with that there passes the colour of his hair and the hanging lower lip of her grandfather and the form of humanity in all its complexity of bones, sinews, nerves, liver and heart, and the form of those pre-human organisms which the embryo will recapitulate in the womb. Behind every spermatozoon lies the whole history of the universe: locked within it lies no inconsiderable part of the world's future. The weight or drive behind it is the momentum of the whole interlocked event which we call Nature up-to-date. And we know now that the 'laws of Nature' cannot supply that momentum. If we believe that God created Nature that momentum comes from Him. The human father is merely an instrument, a carrier, often an unwilling carrier, always simply the last in a long line of carriers—a line that stretches back far beyond his ancestors into pre-human and pre-organic deserts of time, back to the creation of matter itself. That line is in God's hand. It is the instrument by which He normally creates a man. For He is the reality behind both Genius and Venus; no woman ever conceived a child, no mare a foal, without Him. But once, and for a special purpose, He dispensed with that long line which is His instrument: once His life-giving finger touched a woman without passing through the ages of interlocked events. Once the great glove of Nature was taken off His hand. His naked hand touched her. There was of course a unique reason for it. That time He was creating not simply a man but the Man who was to be Himself: was creating Man anew: was beginning, at this divine and human point, the New Creation of all things. The whole soiled and weary universe quivered at this direct injection of essential life—direct, uncontaminated, not drained through all the crowded history of Nature. But it would be out of place here to explore the religious significance of the miracle. We are here concerned with it simply as Miracle—that and nothing more. As far as concerns the creation of Christ's human nature (the Grand Miracle whereby His divine begotten nature enters into it is another matter) the miraculous conception is one more witness that here is Nature's Lord. He is doing now, small and close, what He does in a different fashion for every woman who conceives. He does it this time without a line of human ancestors: but even where He uses human ancestors it is not the less He who gives life.[4] The bed is barren where that great third party, Genius, is not present.

The miracles of *Healing*, to which I turn next, are now in a peculiar position. Men are ready to admit that many of them happened, but are inclined to deny that they were miraculous. The symptoms of very many diseases can be aped by hysteria, and hysteria can often be cured by 'suggestion'. It could, no doubt, be argued that such suggestion is a spiritual power, and therefore (if you like) a supernatural power, and that all instances of 'faith healing' are therefore miracles. But in our terminology they would be miraculous only in the same sense in which every instance of human reason is miraculous: and what we are now looking for is miracles other than that. My own view is that it

[4] Cf. Matthew 23:9.

would be unreasonable to ask a person who has not yet embraced Christianity in its entirety to allow that all the healings mentioned in the Gospels were miracles—that is, that they go beyond the possibilities of human 'suggestions'. It is for the doctors to decide as regards each particular case—supposing that the narratives are sufficiently detailed to allow even probable diagnosis. We have here a good example to what was said in an earlier chapter. So far from belief in miracles depending upon ignorance of natural law, we are here finding for ourselves that ignorance of law makes miracle unascertainable.

Without deciding in detail which of the healings must (apart from acceptance of the Christian faith) be regarded as miraculous, we can however indicate the kind of miracle involved. Its character can easily be obscured by the somewhat magical view which many people still take of ordinary and medical healing. There is a sense in which no doctor ever heals. The doctors themselves would be the first to admit this. The magic is not in the medicine but in the patient's body—in the *vis medicatrix naturae*, the recuperative or self-corrective energy of Nature. What the treatment does is to simulate Natural functions or to remove what hinders them. We speak for convenience of the doctor, or the dressing, healing a cut. But in another sense every cut heals itself: no cut can be healed in a corpse.

That same mysterious force which we call gravitational when it steers the planets and biochemical when it heals a live body, is the efficient cause of all recoveries. And that energy proceeds from God in the first instance. All who are cured are cured by Him, not merely in the sense that His providence provides them with medical assistance and wholesome environments, but also in the sense that their very tissues are repaired by the far-descended energy which, flowing from Him, energises the whole system of Nature. But once He did it visibly to the sick in Palestine, a Man meeting with men. What in its general operations we refer to laws of Nature or once referred to Apollo or Aesculapius thus reveals itself. The Power that always was behind all healings puts on a face and hands. Hence, of course, the apparent chanciness of the miracles. It is idle to complain that He heals those whom He happens to meet, not those whom He doesn't. To be a man means to be in one place and not in another. The world which would not know Him as present everywhere was saved by His becoming *local.*

Christ's single miracle of Destruction, the withering of the fig-tree, has proved troublesome to some people, but I think its significance is plain enough. The miracle is an acted parable, a symbol of God's sentence on all that is 'fruitless' and specially, no doubt, on the official Judaism of that age. That is its moral significance. As a miracle, it again does in focus, repeats small and close, what God does constantly and throughout Nature. We have seen in the previous chapter how God, twisting Satan's weapon out of his hand, had become, since the Fall, the God even of human death. But much more, and perhaps ever since the creation, He has been the God of the death of organisms. In both cases, though in somewhat different ways, He is the God of death because He is the God of Life: the God of human death because through it increase of life now comes—the God of merely organic death because death is part of the very mode by which organic life spreads itself out in Time and yet remains new. A forest a thousand years deep is still collectively alive because some trees are dying and others are growing up. His human face, turned with negation in its eyes upon that one fig-tree, did once what His unincarnate action does to all trees. No tree died that year in Palestine, or any year anywhere, except because God did—or rather ceased to do—something to it.

All the miracles which we have considered so far are Miracles of the Old Creation. In all of them we see the Divine Man focusing for us what the God of Nature has already done on a larger scale. In our next class, the Miracles of Dominion over the Inorganic, we find some that are of the Old Creation and some that are of the New.

When Christ stills the storm He does what God has often done before. God made Nature such that there would be both storms and calms: in that way all storms (except those that are still going on at this moment) have been stilled by God. It is unphilosophical, if you have once accepted the Grand Miracle, to reject the stilling of the storm. There is really no difficulty about adapting the weather conditions of the rest of the world to this one miraculous calm. I myself can still a storm in a room by shutting the window. Nature must make the best she can of it. And to do her justice she makes no trouble at all. The whole system, far from being thrown out of gear (which is what some nervous people seem to think a miracle would do) digests the new situation as easily as an elephant digests a drop of water. She is, as I have said before, an accomplished hostess. But when Christ walks on the water we have a miracle of the New Creation. God had not made the Old Nature, the world before the Incarnation, of such a kind that water would support a human body. This miracle is the foretaste of a Nature that is still in the future. The New creation is just breaking in. For a moment it looks as if it were going to spread. For a moment two men are living in that new world. St Peter also walks on the water—a pace or two: then his trust fails him and he sinks. He is back in Old Nature. That momentary glimpse was a snowdrop of a miracle. The snowdrops show that we have turned the corner of the year. Summer is coming. But it is a long way off and the snowdrops do not last long.

The Miracles of Reversal all belong to the New Creation. It is a Miracle of Reversal when the dead are raised. Old Nature knows nothing of this process: it involves playing backwards a film that we have always seen played forwards. The one or two instances of it in the Gospels are early flowers—what we call spring flowers, because they are prophetic, although they really bloom while it is still winter. And the Miracles of Perfecting or of Glory, the Transfiguration, the Resurrection, and the Ascension, are even more emphatically of the New Creation. These are the true spring, or even the summer, of the world's new year. The Captain, the forerunner, is already in May or June, though His followers on earth are still living in the frosts and east winds of Old Nature—for 'spring comes slowly up this way'.

None of the Miracles of the New Creation can be considered apart from the Resurrection and Ascension: and that will require another chapter.

16

MIRACLES OF THE NEW CREATION

Beware, for fiends in triumph laugh
O'er him who learns the truth by half!
Beware; for God will not endure
For men to make their hope more pure
Than His good promise, or require
Another than the five-stringed lyre[1]
Which He has vowed again to the hands
Devout of him who understands
To tune it justly here!

C. PATMORE, *The Victories of Love*

In the earliest days of Christianity an 'apostle' was first and foremost a man who claimed to be an eyewitness of the Resurrection. Only a few days after the Crucifixion when two candidates were nominated for the vacancy created by

[1] i.e. the Body with its five senses.

the treachery of Judas, their qualification was that they had known Jesus personally both before and after His death and could offer first-hand evidence of the Resurrection in addressing the outer world (Acts 1:22). A few days later St Peter, preaching the first Christian sermon, makes the same claim—'God raised Jesus, of which we all (we Christians) are witnesses (Acts 2:32). In the first Letter to the Corinthians, St Paul bases his claim to apostleship on the same ground—'Am I not an apostle? Have I not seen the Lord Jesus?' (1:9).

As this qualification suggests, to preach Christianity meant primarily to preach the Resurrection. Thus people who had heard only fragments of St Paul's teaching at Athens got the impression that he was talking about two new gods, Jesus and Anastasis (i.e. Resurrection) (Acts 17:18). The Resurrection is the central theme in every Christian sermon reported in the Acts. The Resurrection, and its consequences, were the 'gospel' or good news which the Christians brought: what we call the 'gospels', the narratives of Our Lord's life and death, were composed later for the benefit of those who had already accepted the *gospel*. They were in no sense the basis of Christianity: they were written for those already converted. The miracle of the Resurrection, and the theology of that miracle, comes first: the biography comes later as a comment on it.

Nothing could be more unhistorical than to pick out selected sayings of Christ from the gospels and to regard those as the datum and the rest of the New Testament as a construction upon it. The first fact in the history of Christendom is a number of people who say they have seen the Resurrection. If they had died without making anyone else believe this 'gospel' no gospels would ever have been written.

It is very important to be clear about what these people meant. When modern writers talk of the Resurrection they usually mean one particular moment—the discovery of the Empty Tomb and the appearance of Jesus a few yards away from it. The story of that moment is what Christian apologists now chiefly try to support and sceptics chiefly try to impugn. But this almost exclusive concentration on the first five minutes or so of the Resurrection would have astonished the earliest Christian teachers. In claiming to have seen the Resurrection they were not necessarily claiming to have seen *that*. Some of them had, some of them had not. It had no more importance than any of the other appearances of the risen Jesus—apart from the poetic and dramatic importance which the beginnings of things must always have. What they were claiming was that they had all, at one time or another, met Jesus during the six or seven weeks that followed His death. Sometimes they seem to have been alone when they did so, but on one occasion twelve of them saw Him together, and on another occasion about five hundred of them. St Paul says that the majority of the five hundred were still alive when he wrote the First Letter to the Corinthians, i.e. in about 55 AD.

The 'Resurrection' to which they bore witness was, in fact, not the action of rising from the dead but the state of having risen; a state, as they held, attested by intermittent meetings during a limited period (except for the special, and in some ways different, meeting vouchsafed to St Paul). This termination of the period is important, for, as we shall see, there is no possibility of isolating the doctrine of the Resurrection from that of the Ascension.

The next point to notice is that the Resurrection was not regarded simply or chiefly as evidence for the immortality of the soul. It is, of course, often so regarded today: I have heard a man maintain that 'the importance of the Resurrection is that it proves *survival*'. Such a view cannot at any point be reconciled with the language of the New Testament. On such a view Christ would simply have done what all men do when they die: the only novelty would have been that in His case we were allowed to see it happening. But there is not in Scripture the faintest suggestion that the Resurrection was new evidence for something that had *in fact* been always happening. The New Testament writers speak as if Christ's achievement in rising from the dead was the first event of its kind in the whole history of the universe. He is the 'first fruits', the 'pioneer of life'. He has forced open a door that has been locked since the death of the first man. He has met, fought, and beaten the King of Death. Everything is different because He has done so. This is the beginning of the New Creation: a new chapter in cosmic history has opened.

I do not mean, of course, that the writers of the New Testament disbelieved in 'survival'. On the contrary they believed in it so readily that Jesus on more than one occasion had to assure them that He was not a ghost. From the earliest times the Jews, like many other nations, had believed that man possessed a 'soul' or *Nephesh* separable from the body, which went at death into the shadowy world called *Sheol:* a land of forgetfulness and imbecility where none called upon Jehovah any more, a land half unreal and melancholy like the Hades of the Greeks or the Niflheim of the Norsemen. From it shades could return and appear to the living, as Samuel's shade had done at the command of the Witch of Endor. In much more recent times there had arisen a more cheerful belief that the righteous passed at death to 'heaven'. Both doctrines are doctrines of 'the immortality of the soul' as a Greek or modern Englishman understands it: and both are quite irrelevant to the story of the Resurrection. The writers look upon this event as an absolute novelty. Quite clearly they do not think they have been haunted by a ghost from Sheol, nor even that they have had a vision of a 'soul' in 'heaven'. It must be clearly understood that if the Psychical Researchers succeeded in proving 'survival' and showed that the Resurrection was an instance of it, they would not be supporting the Christian faith but refuting it. If that were all that had happened the original 'gospel' would have been untrue. What the apostles claimed to have seen did not corroborate, nor exclude, and had indeed nothing to do with, either the doctrine of 'heaven' or the doctrine of Sheol. Insofar as it corroborated anything it corroborated a third Jewish belief which is quite distinct from both these. This third doctrine taught that in 'the day of Jahweh' peace would be restored and world dominion given to Israel under a righteous King: and that when this happened the righteous dead, or some of them, would come back to earth—not as floating wraiths but as solid men who cast shadows in the sunlight and made a noise when they tramped the floors. 'Awake and sing, ye that dwell in the dust', said Isaiah, 'And the earth shall cast out the dead' (26:19). What the apostles thought they had seen was, if not that, at any rate a lonely first instance of that: the first movement of a great wheel beginning to turn in the direction opposite to that which all men hitherto had observed. Of all the ideas entertained by man about death it is this one, and this one only, which the story of the Resurrection tends to confirm. If the story is false then it is this Hebrew myth of resurrection which begot it. If the story is true then the hint and anticipation of the truth is to be found not in popular ideas about ghosts nor in eastern doctrines of reincarnation nor in philosophical speculations about the immortality of the soul, but exclusively in the Hebrew prophecies of the return, the restoration, the great reversal. Immortality simply as immortality is irrelevant to the Christian claim.

There are, I allow, certain respects in which the risen Christ resembles the 'ghost' of popular tradition. Like a ghost He 'appears' and 'disappears': locked doors are no obstacle to Him. On the other hand He Himself vigorously asserts that He is corporeal (Luke 24: 39-40) and eats broiled fish. It is at this point that the modern reader becomes uncomfortable. He becomes more uncomfortable still at the word, 'Don't touch me; I have not yet gone up to the Father' (John 20:17). For voices and apparitions we are, in some measure, prepared. But what is this that must not be touched? What is all this about going 'up' to the Father? Is He not already 'with the Father' in the only sense that matters? What can 'going up' be except a metaphor for *that?* And if so, why has He 'not yet' gone? These discomforts arise because the story the 'apostles' actually had to tell begins at this point to conflict with the story we expect and are determined beforehand to read into their narrative.

We expect them to tell of a risen life which is purely 'spiritual' in the negative sense of that word: that is, we use the word 'spiritual' to mean not what it is but what it is not. We mean a life without space, without history, without environment, with no sensuous elements in it. We also, in our heart of hearts, tend to slur over the risen *manhood* of Jesus, to conceive Him, after death, simply returning into Deity, so that the Resurrection would be no more than the reversal or undoing of the Incarnation. That being so, all references to the risen *body* make us uneasy: they raise awkward questions. For as long as we hold the negatively spiritual view, we have not really been believing in that body at all. We have thought (whether we acknowledged it or not) that the body was not objective: that it was an appearance sent by God to assure the disciples of truths otherwise incommunicable. But what truths? If the truth is that after death there comes a negatively spiritual life, an eternity of mystical experience, what more misleading way of communicating it could possibly be found than the appearance of a human form which eats broiled fish? Again, on such a view, the body would really be a hallucination. And any theory of hallucination breaks down on the fact (and if it is invention it is the oddest invention that ever entered the mind of man) that on three separate occasions this hallucination was not immediately recognised as Jesus (Luke 24:13-31; John 20:15, 21:4). Even granting that God sent a holy hallucination to teach truths already widely believed without it, and far more easily taught by other methods, and certain to be completely obscured by this, might we not at least hope that He would get the face of the hallucination *right?* Is He who made all faces such a bungler that He cannot even work up a recognisable likeness of the Man who was Himself?

It is at this point that awe and trembling fall upon us as we read the records. If the story is false, it is at least a much stranger story than we expected, something for which philosophical 'religion', psychical research, and popular superstition have all alike failed to prepare us. If the story is true, then a wholly new mode of being has arisen in the universe.

The body which lives in that new mode is like, and yet unlike, the body His friends knew before the execution. It is differently related to space and probably to time, but by no means cut off from all relation to them. It can perform the animal act of eating. It is so related to matter, as we know it, that it can be touched, though at first it had better not be touched. It has also a history before it which is in view from the first moment of the Resurrection; it is presently going to become different or go somewhere else. That is why the story of the Ascension cannot be separated from that of the Resurrection. All the accounts suggest that the appearances of the Risen Body came to an end; some describe an abrupt end about six weeks after the death. And they describe this abrupt end in a way which presents greater difficulties to the modern mind than any other part of Scripture. For here, surely, we get the implication of all those primitive crudities to which I have said that Christians are not committed: the vertical ascent like a balloon, the local Heaven, the decorated chair to the right of the Father's throne. 'He was caught up into the sky (*ouranos*)', says St Mark's Gospel 'and sat down at the right hand of God'. 'He was lifted up', says the author of Acts 'and a cloud cut Him off from their sight'.

It is true that if we wish to get rid of these embarrassing passages we have the means to do so. The Marcan one probably formed no part of the earliest text of St Mark's Gospel: and you may add that the Ascension, though constantly implied throughout the New Testament, is described only in these two places. Can we then simply drop the Ascension story? The answer is that we can do so only if we regard the Resurrection appearances as those of a ghost or hallucination. For a phantom can just fade away; but an objective entity must go somewhere—something must happen to it. And if the Risen Body were not objective, then all of us (Christian or not) must invent some explanation for the disappearance of the corpse. And all Christians must explain why God sent or permitted a 'vision' or 'ghost' whose behaviour seems almost exclusively directed to convincing the disciples that it was not a vision or a ghost but a really corporeal being. If it were a vision then it was the most systematically deceptive and lying vision on record. But if it were real, then something happened to it after it ceased to appear. You cannot take away the Ascension without putting something else in its place.

The records represent Christ as passing after death (as no man had passed before) neither into a purely, that is, negatively, 'spiritual' mode of existence nor into a 'natural' life such as we know, but into a life which has its own, new Nature. It represents Him as withdrawing six weeks later, into some different mode of existence. It says—He says—that He goes 'to prepare a place for us'. This presumably means that He is about to create that whole new Nature which will provide the environment or conditions for His glorified humanity and, in Him, for ours. The picture is not what we expected—though whether it is less or more probable and philosophical on that account is another question. It is not the picture of an escape from any and every kind of Nature into some unconditioned and utterly transcendent life. It is the picture of a new human nature, and a new Nature in general, being brought into existence. We must, indeed, believe the risen body to be extremely different from the mortal body: but the existence, in that new state, of anything that could in any sense be described as 'body' at all, involves some sort of spatial relations and in the long run a whole new universe. That is the picture—not of unmaking but of remaking. The old field of space, time, matter, and the senses is to be weeded, dug, and sown for a new crop. We may be tired of that old field: God is not.

And yet the very way in which this New Nature begins to shine in has a certain affinity with the habits of Old Nature. In Nature as we know her, things tend to be anticipated. Nature is fond of 'false dawns', of precursors: thus, as I said before, some flowers come before true spring: sub-men (the evolutionists would have it) before the true men. So, here also, we get Law before Gospel, animal sacrifices foreshadowing the great sacrifice of God to God, the Baptist before the Messiah, and those 'miracles of the New Creation' which come before the Resurrection. Christ's walking on the water, and His raising of Lazarus fall in this class. Both give us hints of what the New Nature will be like.

In the Walking on the Water we see the relations of spirit and Nature so altered that Nature can be made to do whatever spirit pleases. This new obedience of Nature is, of course, not to be separated even in thought from spirit's own obedience to the Father of Spirits. Apart from that proviso such obedience by Nature, if it were possible, would result in chaos: the evil dream of Magic arises from finite spirit's longing to get that power without paying that price. The evil reality of lawless applied science (which is Magic's son and heir) is actually reducing large tracts of Nature to disorder and sterility at this very moment. I do not know how radically Nature herself would need to be altered to make her thus obedient to spirits, when spirits have become wholly obedient to their source. One thing at least we must observe. If we are in fact spirits, not Nature's offspring, then there must be some point (probably the brain) at which created spirit even now can produce effects on matter not by manipulation or technics but simply by the wish to do so. If that is what you mean by Magic then Magic is a reality manifested every time you move your hand or think a thought. And Nature, as we have seen, is not destroyed but rather perfected by her servitude.

The raising of Lazarus differs from the Resurrection of Christ Himself because Lazarus, so far as we know, was not raised to a new and more glorious mode of existence but merely restored to the sort of life he had had before. The fitness of the miracle lies in the fact that He who will raise all men at the general resurrection here does it small and close, and in an inferior—a merely anticipatory—fashion. For the mere restoration of Lazarus is as inferior in splendour to the *glorious* resurrection of the New Humanity as stone jars are to the green and growing vine or five little barley loaves to all the waving bronze and gold of a fat valley ripe for harvest. The resuscitation of Lazarus, so far as we can see, is simple reversal: a series of changes working in the direction opposite to that we have always experienced. At death, matter which has been organic, begins to flow away into the inorganic, to be finally scattered and used (some of it) by other organisms. The resurrection of Lazarus involves the reverse process. The general resurrection involves the reverse process universalised—a rush of matter toward organisation at the call of spirits which require it. It is presumably a foolish fancy (not justified by the words of Scripture) that each spirit should recover those particular units of matter which he ruled before. For one thing, they would not be enough to go round: we all live in second-hand suits and there are doubtless atoms in my chin which have served many another man, many a dog, many an eel, many a dinosaur. Nor does the unity of our bodies, even in this present life, consist in retaining the same particles. My form remains one, though the matter in it changes continually. I am, in that respect, like a curve in a waterfall.

But the miracle of Lazarus, though only anticipatory in one sense, belongs emphatically to the New Creation, for nothing is more definitely excluded by Old Nature than any return to a *status quo*. The pattern of Death and Rebirth never restores the previous individual organism. And similarly, on the inorganic level, we are told that Nature never restores order where disorder has once occurred. 'Shuffling,' said Professor Eddington, 'is the thing Nature never undoes'. Hence we live in a universe where organisms are always getting more disordered. These laws between them—irreversible death and irreversible entropy—cover almost the whole of what St Paul calls the 'vanity' of Nature: her futility, her ruinousness. And the film is never reversed. The movement from more order to less almost serves to determine the direction in which Time is flowing. You could almost define the future as the period in which what is now living will be dead and in which what order still remains will be diminished.

But entropy by its very character assures us that though it may be the universal rule in the Nature we know, it cannot be universal absolutely. If a man says 'Humpty Dumpty is falling,' you see at once that this is not a complete story.

The bit you have been told implies both a later chapter in which Humpty Dumpty will have reached the ground, and an earlier chapter in which he was still seated on the wall. A Nature which is 'running down' cannot be the whole story. A clock can't run down unless it has been wound up. Humpty Dumpty can't fall off a wall which never existed. If a Nature which disintegrates order were the whole of reality, where would she find any order to disintegrate? Thus on any view there must have been a time when processes the reverse of those we now see were going on: a time of winding up. The Christian claim is that those days are not gone for ever. Humpty Dumpty is going to be replaced on the wall—at least in the sense that what has died is going to recover life, probably in the sense that the inorganic universe is going to be reordered. Either Humpty Dumpty will never reach the ground (being caught in mid-fall by the everlasting arms) or else when he reaches it he will be put together again and replaced on a new and better wall. Admittedly, science discerns no 'king's horses and men' who can 'put Humpty Dumpty together again'. But you would not expect her to. She is based on observation: and all our observations are observations of Humpty Dumpty in mid-air. They do not reach either the wall above or the ground below—much less the King with his horses and men hastening towards the spot.

The Transfiguration or 'Metamorphosis' of Jesus is also, no doubt, an anticipatory glimpse of something to come. He is seen conversing with two of the ancient dead. The change which His own human form had undergone is described as one to luminosity, to 'shining whiteness'. A similar whiteness characterises His appearance at the beginning of the book of Revelation. One rather curious detail is that this shining or whiteness affected His clothes as much as His body. St Mark indeed mentions the clothes more explicitly than the face, and adds, with his inimitable naïvety, that 'no laundry could do anything like it'. Taken by itself this episode bears all the marks of a 'vision': that is, of an experience which, though it may be divinely sent and may reveal great truth, yet is not, objectively speaking, the experience it seems to be. But if the theory of 'vision' (or holy hallucination) will not cover the Resurrection appearances, it would be only a multiplying of hypotheses to introduce it here. We do not know to what phase or feature of the New Creation this episode points. It may reveal some special glorifying of Christ's manhood at some phase of its history (since history it apparently has) or it may reveal the glory which that manhood always has in its New Creation: it may even reveal a glory which all risen men will inherit. We do not know.

It must indeed be emphasised throughout that we know and can know very little about the New Nature. The task of the imagination here is not to forecast it but simply, by brooding on many possibilities, to make room for a more complete and circumspect agnosticism. It is useful to remember that even now senses responsive to different vibrations would admit us to quite new worlds of experience: that a multi-dimensional space would be different, almost beyond recognition, from the space we are now aware of, yet not discontinuous from it: that time may not always be for us, as it now is, unilinear and irreversible: that others parts of Nature might some day obey us as our cortex now does. It is useful not because we can trust these fancies to give us any positive truths about the New Creation but because they teach us not to limit, in our rashness, the vigour and variety of the new crops which this old field might yet produce. We are therefore compelled to believe that nearly all we are told about the New Creation is metaphorical. But not quite all. That is just where the story of the Resurrection suddenly jerks us back like a tether. The local appearances, the eating, the touching, the claim to be corporeal, must be either reality or sheer illusion. The New Nature is, in the most troublesome way, interlocked at some points with the Old. Because of its novelty we have to think of it, for the most part, metaphorically: but because of the partial interlocking, some facts about it come through into our present experience in all their literal facthood—just as some facts about an organism are inorganic facts, and some facts about a solid body are facts of linear geometry.

Even apart from that, the mere idea of a New Nature, a Nature beyond Nature, a systematic and diversified reality which is 'supernatural' in relation to the world of our five present senses but 'natural' from its own point of view, is profoundly shocking to a certain philosophical preconception from which we all suffer. I think Kant is at the root of it. It may be expressed by saying that we are prepared to believe either in a reality with one floor or in a reality with two floors, but not in a reality like a skyscraper with several floors. We are prepared, on the one hand, for the sort of reality that Naturalists believe in. That is a one-floor reality: this present Nature is all that there is. We are also prepared for reality as 'religion' conceives it: a reality with a ground floor (Nature) and then above that one other floor and one only—an eternal, spaceless, timeless, spiritual Something of which we can have no images and which, if it presents itself to human consciousness at all, does so in a mystical experience which shatters all our categories of thought. What we are not prepared for is anything in between. We feel quite sure that the first step beyond the world of our present experience must lead either nowhere at all or else into the blinding abyss of undifferentiated spirituality, the unconditioned, the absolute. That is why many believe in God who cannot believe in angels and an angelic world. That is why many believe in immortality who cannot believe in the resurrection of the body. That is why Pantheism is more popular than Christianity, and why many desire a Christianity stripped of its miracles. I cannot now understand, but I well remember, the passionate conviction with which I myself once defended this prejudice. Any rumour of floors or levels intermediate between the Unconditioned and the world revealed by our present senses I rejected without trial as 'mythology'.

Yet it is very difficult to see any rational grounds for the dogma that reality must have no more than two levels. There cannot, from the nature of the case, be evidence that God never created and never will create, more than one system. Each of them would be at least extra-natural in relation to all the others: and if any of them is more concrete, more permanent, more excellent, and richer than another it will be to that other *super*-natural. Nor will a partial contact between any two obliterate their distinctness. In that way there might be Natures piled upon Natures to any height God pleased, each Supernatural to that below it and Subnatural to that which surpassed it. But the tenor of Christian teaching is that we are actually living in a situation even more complex than that. A new Nature is being not merely made but made out of an old one. We live amid all the anomalies, inconveniences, hopes, and excitements of a house that is being rebuilt. Something is being pulled down and something going up in its place.

To accept the idea of intermediate floors—which the Christian story will, quite simply, force us to do if it is not a falsehood—does not of course involve losing our spiritual apprehension of the top floor of all. Most certainly, beyond all worlds, unconditioned and unimaginable, transcending discursive thought, there yawns for ever the ultimate Fact, the fountain of all other facthood, the burning and undimensioned depth of the Divine Life. Most certainly also, to be united with that Life in the eternal Sonship of Christ is, strictly speaking, the only thing worth a moment's consideration. And in so far as *that* is what you mean by *Heaven,* Christ's divine Nature never left it, and therefore never returned to it: and His human nature ascended thither not at the moment of the Ascension but at every moment. In that sense not one word that the spiritualisers have uttered will, please God, ever be unsaid by me. But it by no means follows that there are not other truths as well. I allow, indeed I insist, that Christ cannot be at 'the right hand of God' except in a metaphorical sense. I allow and insist that the Eternal Word, the Second Person of the Trinity, can never be, nor have been, confined to any place at all: it is rather in Him that all places exist. But the records say that the glorified, but still in some sense corporeal, Christ withdrew into some different mode of being about six weeks after the Crucifixion: and that He is 'preparing a place' for us. The statement in St Mark that He sat down at the right hand of God we must take as a metaphor: it was indeed, even for the writer, a poetical quotation, from Psalm 110. But the statement that the holy Shape went up and vanished does not permit the same treatment.

What troubles us here is not simply the statement itself but what (we feel sure) the author meant by it. Granted that there are different Natures, different levels of being, distinct but not always discontinuous—granted that Christ withdrew from one of these to another, that His withdrawal from one was indeed the first step in His creation of the other—what precisely should we expect the onlookers to see? Perhaps mere instantaneous vanishing would make us most comfortable. A sudden break between the perceptible and the imperceptible would worry us less than any kind of joint. But if the spectators say they saw first a short vertical movement and then a vague luminosity (that is what 'cloud' presumably means here as it certainly does in the account of the Transfiguration) and then nothing—

have we any reason to object? We are well aware that increased distance from the centre of this planet could not *in itself* be equated with increase of power or beatitude. But this is only saying that *if* the movement had no connection with such spiritual events, why then it had no connection with them.

Movement (in any direction but one) away from the position momentarily occupied by our moving Earth will certainly be to us movement 'upwards'. To say that Christ's passage to a new 'Nature' could involve no such movement, or no movement at all, within the 'Nature' he was leaving, is very arbitrary. Where there is passage, there is departure; and departure is an event in the region from which the traveller is departing. All this, even on the assumption that the Ascending Christ is in a three-dimensional space. If it is not that kind of body, and space is not that kind of space, then we are even less qualified to say what the spectators of this entirely new event might or might not see or feel as if they had seen. There is, of course, no question of a human body as we know it existing in interstellar space as we know it. The Ascension belongs to a New Nature. We are discussing only what the 'joint' between the Old Nature and the new, the precise moment of transition, would look like.

But what really worries us is the conviction that, whatever we say, the New Testament writers meant something quite different. We feel sure that they thought they had seen their Master setting off on a journey for a local 'Heaven' where God sat in a throne and where there was another throne waiting for Him. And I believe that in a sense that is just what they did think. And I believe that, for this reason, whatever they had actually seen (sense perception, almost by hypothesis, would be confused at such a moment) they would almost certainly have remembered it as a vertical movement. What we must not say is that they 'mistook' local 'Heavens' and celestial throne-rooms and the like for the 'spiritual' Heaven of union with God and supreme power and beatitude. You and I have been gradually disentangling different senses of the word *Heaven* throughout this chapter. It may be convenient here to make a list. *Heaven* can mean (1) The unconditioned Divine Life beyond all worlds. (2) Blessed participation in that Life by a created spirit. (3) The whole Nature or system of conditions in which redeemed human spirits, still remaining human, can enjoy such participation fully and for ever. This is the Heaven Christ goes to 'prepare' for us. (4) The physical Heaven, the sky, the space in which Earth moves. What enables us to distinguish these senses and hold them clearly apart is not any special spiritual purity but the fact that we are the heirs to centuries of logical analysis: not that we are sons to Abraham but that we are sons to Aristotle. We are not to suppose that the writers of the New Testament mistook Heaven in sense four or three for Heaven in sense two or one. You cannot mistake a half sovereign for a sixpence until you know the English system of coinage—that is, until you know the difference between them. In their idea of Heaven all these meanings were latent, ready to be brought out by later analysis. They never thought merely of the blue sky or merely of a 'spiritual' heaven. When they looked up at the blue sky they never doubted that there, whence light and heat and the precious rain descended, was the home of God: but on the other hand, when they thought of one ascending to that Heaven they never doubted He was 'ascending' in what we should call a 'spiritual' sense. The real and pernicious period of literalism comes far later, in the Middle Ages and the seventeenth century, when the distinctions have been made and heavy-handed people try to force the separated concepts together again in wrong ways. The fact that Galilean shepherds could not distinguish what they saw at the Ascension from that kind of ascent which, by its very nature, could never be seen at all, does not prove on the one hand that they were unspiritual, nor on the other that they saw nothing. A man who really believes that 'Heaven' is in the sky may well, in his heart, have a far truer and more spiritual conception of it than many a modern logician who could expose that fallacy with a few strokes of his pen. For he who does the will of the Father shall know the doctrine. Irrelevant material splendours in such a man's idea of the vision of God will do no harm, for they are not there for their own sakes. Purity from such images in a merely theoretical Christian's idea will do no good if they have been banished only by logical criticism.

But we must go a little further than this. It is not an accident that simple-minded people, however spiritual, should blend the ideas of God and Heaven and the blue sky. It is a fact, not a fiction, that light and life-giving heat do come down from the sky to Earth. The analogy of the sky's role to begetting and of the Earth's role to bearing is sound as far as it goes. The huge dome of the sky is of all things sensuously perceived the most like infinity. And when God made space and worlds that move in space, and clothed our world with air, and gave us such eyes and such imaginations as those we have, He knew what the sky would mean to us. And since nothing in His work is accidental, if He knew, He intended. We cannot be certain that this was not indeed one of the chief purposes for which Nature was created; still less that it was not one of the chief reasons why the withdrawal was allowed to affect human senses as a movement upwards. (A disappearance into the Earth would beget a wholly different religion.) The ancients in letting the spiritual symbolism of the sky flow straight into their minds without stopping to discover by analysis that it was a symbol, were not entirely mistaken. In one way they were perhaps less mistaken than we.

For we have fallen into an opposite difficulty. Let us confess that probably every Christian now alive finds a difficulty in reconciling the two things he has been told about 'heaven'—that it is, on the one hand, a life in Christ, a vision of God, a ceaseless adoration, and that it is, on the other hand, a bodily life. When we seem nearest to the vision of God in this life, the body seems almost an irrelevance. And if we try to conceive our eternal life as one in a body (any kind of body) we tend to find that some vague dream of Platonic paradises and gardens of the Hesperides has substituted itself for that mystical approach which we feel (and I think rightly) to be more important. But if that discrepancy were final then it would follow—which is absurd—that God was originally mistaken when He introduced our spirits into the Natural order at all. We must conclude that the discrepancy itself is precisely one of the disorders which the New Creation comes to heal. The fact that the body, and locality and locomotion and time, now feel irrelevant to the highest reaches of the spiritual life is (like the fact that we can think of our bodies as 'coarse') a *symptom*. Spirit and Nature have quarrelled in us; that is our disease. Nothing we can yet do enables us to imagine its complete healing. Some glimpses and faint hints we have: in the Sacraments, in the use made of sensuous imagery by the great poets, in the best instances of sexual love, in our experiences of the earth's beauty. But the full healing is utterly beyond our present conceptions. Mystics have got as far in contemplation of God as the point at which the senses are banished: the further point, at which they will be put back again, has (to the best of my knowledge) been reached by no one. The destiny of redeemed man is not less but more unimaginable than mysticism would lead us to suppose—because it is full of semi-imaginables which we cannot at present admit without destroying its essential character.

One point must be touched on because, though I kept silence, it would none the less be present in most readers' minds. The letter and spirit of scripture, and of all Christianity, forbid us to suppose that life in the New Creation will be a sexual life; and this reduces our imagination to the withering alternative either of bodies which are hardly recognisable as human bodies at all or else of a perpetual fast. As regards the fast, I think our present outlook might be like that of a small boy who, on being told that the sexual act was the highest bodily pleasure should immediately ask whether you ate chocolates at the same time. On receiving the answer 'No,' he might regard absence of chocolates as the chief characteristic of sexuality. In vain would you tell him that the reason why lovers in their carnal raptures don't bother about chocolates is that they have something better to think of. The boy knows chocolate: he does not know the positive thing that excludes it. We are in the same position. We know the sexual life; we do not know, except in glimpses, the other thing which, in Heaven, will leave no room for it. Hence where fullness awaits us we anticipate fasting. In denying that sexual life, as we now understand it, makes any part of the final beatitude, it is not of course necessary to suppose that the distinction of sexes will disappear. What is no longer needed for biological purposes may be expected to survive for splendour. Sexuality is the instrument both of virginity and of conjugal virtue; neither men nor women will be asked to throw away weapons they have used victoriously. It is the beaten and the fugitives who throw away their swords. The conquerors sheathe theirs and retain them. 'Trans-sexual' would be a better word than 'sexless' for the heavenly life.

I am well aware that this last paragraph may seem to many readers unfortunate and to some comic. But that very comedy, as I must repeatedly insist, is the symptom of our estrangement, as spirits, from Nature and our estrangement, as animals, from Spirit. The whole conception of the

New Creation involves the belief that this estrangement will be healed. A curious consequence will follow. The archaic type of thought which could not clearly distinguish spiritual 'Heaven' from the sky, is from our point of view a confused type of thought. But it also resembles and anticipates a type of thought which will one day be true. That archaic sort of thinking will become simply the correct sort when Nature and Spirit are fully harmonised—when Spirit rides Nature so perfectly that the two together make rather a *Centaur* than a mounted knight. I do not mean necessarily that the blending of Heaven and sky, in particular, will turn out to be specially true, but that that kind of blending will accurately mirror the reality which will then exist. There will be no room to get the finest razor-blade of thought in between Spirit and Nature. Every state of affairs in the New Nature will be the perfect expression of a spiritual state and every spiritual state the perfect informing of, and bloom upon, a state of affairs; one with it as the perfume with a flower or the 'spirit' of great poetry with its form. There is thus in the history of human thought, as elsewhere, a pattern of death and rebirth. The old, richly imaginative thought which still survives in Plato has to submit to the deathlike, but indispensable, process of logical analysis: nature and spirit, matter and mind, fact and myth, the literal and the metaphorical, have to be more and more sharply separated, till at last a purely mathematical universe and a purely subjective mind confront one another across an unbridgeable chasm. But from this descent also, if thought itself is to survive, there must be reascent and the Christian conception provides for it. Those who attain the glorious resurrection will see the dry bones clothed again with flesh, the fact and the myth remarried, the literal and the metaphorical rushing together.

The remark so often made that 'Heaven is a state of mind' bears witness to the wintry and deathlike phase of this process in which we are now living. The implication is that if Heaven is a state of mind—or, more correctly, of the spirit—then it must be only a state of the spirit, or at least that anything else, if added to that state of spirit, would be irrelevant. That is what every great religion *except* Christianity would say. But Christian teaching by saying that God made the world and called it good teaches that Nature or environment cannot be simply irrelevant to spiritual beatitude in general, however far in one particular Nature, during the days of her bondage, they may have drawn apart. By teaching the resurrection of the body it teaches that Heaven is not merely a state of the spirit but a state of the body as well: and therefore a state of Nature as a whole. Christ, it is true, told His hearers that the

Kingdom of Heaven was 'within' or 'among' them. But His hearers were not *merely* in 'a state of mind'. The planet He had created was beneath their feet, His sun above their heads; blood and lungs and guts were working in the bodies He had invented, photons and sound waves of His devising were blessing them with the sight of His human face and the sound of His voice. We are never *merely* in a state of mind. The prayer and the meditation made in howling wind or quiet sunshine, in morning alacrity or evening resignation, in youth or age, good health or ill, may be equally, but are differently, blessed. Already in this present life we have all seen how God can take up all these seeming irrelevances into the spiritual fact and cause them to bear no small part in making the blessing of that moment to be the particular blessing it was—as fire can burn coal and wood equally but a wood fire is different from a coal one. From this factor of environment Christianity does not teach us to desire a total release. We desire, like St Paul, not to be unclothed but to be re-clothed: to find not the formless Everywhere-and-Nowhere but the promised land, that Nature which will be always and perfectly—as present Nature is partially and intermittently—the instrument for that music which will then arise between Christ and us.

And what, you ask, does it matter? Do not such ideas only excite us and distract us from the more immediate and more certain things, the love of God and our neighbours, the bearing of the daily cross? If you find that they so distract you, think of them no more. I most fully allow that it is of more importance for you or me today to refrain from one sneer or to extend one charitable thought to an enemy than to know all that angels and archangels know about the mysteries of the New Creation. I write of these things not because they are the most important but because this book is about miracles. From the title you cannot have expected a book of devotion or of ascetic theology. Yet I will not admit that the things we have been discussing for the last few pages are of no importance for the practice of the Christian life. For I suspect that our conception of Heaven as *merely* a state of mind is not unconnected with the fact that the specifically Christian virtue of Hope has in our time grown so languid. Where our fathers, peering into the future, saw gleams of gold, we see only the mist, white, featureless, cold and never moving.

The thought at the back of all this negative spirituality is really one forbidden to Christians. They, of all men, must not conceive spiritual joy and worth as things that need to be rescued or tenderly protected from time and place and matter and the senses. Their God is the God of corn and oil and wine. He is the glad Creator. He has become Himself incarnate. The sacraments have been instituted. Certain spiritual gifts are offered us only on condition that we perform certain bodily acts. After that we cannot really be in doubt of His intention. To shrink back from all that can be called Nature into negative spirituality is as if we ran away from horses instead of learning to ride. There is in our present pilgrim condition plenty of room (more room than most of us like) for abstinence and renunciation and mortifying our natural desires. But behind all asceticism the thought should be, 'Who will trust us with the true wealth if we cannot be trusted even with the wealth that perishes?' Who will trust me with a spiritual body if I cannot control even an earthly body? These small and perishable bodies we now have were given to us as ponies are given to schoolboys. We must learn to manage: not that we may some day be free of horses altogether but that some day we may ride bare-back, confident and rejoicing, those greater mounts, those winged, shining and world-shaking horses which perhaps even now expect us with impatience, pawing and snorting in the King's stables. Not that the gallop would be of any value unless it were a gallop with the King; but how else—since He has retained His own charger—should we accompany Him?

17

EPILOGUE

> If you leave a thing alone you leave it to a torrent of change. If you leave a white post alone it will soon be a black post.
>
> G. K. CHESTERTON, *Orthodoxy*

My work ends here. If, after reading it, you now turn to study the historical evidence for yourself, begin with the New Testament and not with the books about it. If you do not know Greek get it in a modern translation. Moffat's is probably the best: Monsignor Knox is also good. I do not advise the *Basic English* version. And when you turn from the New Testament to modern scholars, remember that you go among them as a sheep among wolves. Naturalistic assumptions, beggings of the question such as that which I noted on the first page of this book, will meet you on every side—even from the pens of clergymen. This does not mean (as I was once tempted to suspect) that these clergymen are disguised apostates who deliberately exploit the position and the livelihood given them by the Christian Church to undermine Christianity. It comes partly from what we may call a 'hangover'. We all have Naturalism in our bones and even conversion does not at once work the infection out of our system. Its assumptions rush back upon the mind the moment vigilance is relaxed. And in part the procedure of these scholars arises from the feeling which is greatly to their credit—which indeed is honourable to the point of being Quixotic. They are anxious to allow to the enemy every advantage he can with any show of fairness claim. They thus make it part of their method to eliminate the supernatural wherever it is even remotely possible to do so, to strain natural explanation even to the breaking point before they admit the least suggestion of miracle. Just in the same spirit some examiners tend to overmark any candidate whose opinions and character, as revealed by his work, are revolting to them. We are so afraid of being led into unfairness by our instant dislike of the man that we are liable to overshoot the mark and treat him too kindly. Many modern Christian scholars overshoot the mark for a similar reason.

In using the books of such people you must therefore be continually on guard. You must develop a nose like a bloodhound for those steps in the argument which depend not on historical and linguistic knowledge but on the concealed assumption that miracles are impossible, improbable or improper. And this means that you must really re-educate yourself: must work hard and consistently to eradicate from your mind the whole type of thought in which we have all been brought up. It is the type of thought which, under various disguises, has been our adversary throughout this book. It is technically called *Monism*; but perhaps the unlearned reader will understand me best if I call it *Everythingism*. I mean by this the belief that 'everything', or 'the whole show', must be self-existent, must be more important than every particular thing, and must contain all particular things in such a way that they cannot be really very different from one another—that they must be not merely 'at one', but one. Thus the Everythingist, if he starts from God, becomes a Pantheist; there must be nothing that is not God. If he starts from Nature he becomes a Naturalist; there must be nothing that is not Nature. He thinks that everything is in the long run 'merely' a precursor or a development or a relic or an instance or a disguise, of everything else. This philosophy I believe to be profoundly untrue. One of the moderns has said that reality is 'incorrigibly plural'. I think he is right. All things come from One. All things are related—related in different and complicated ways. But all things are not one. The word 'everything' should mean simply the total (a total to be reached, if we knew enough, by enumeration) of all the things that exist at a given moment. It must not be given a mental capital letter; must not (under the influence of picture thinking) be turned into a sort of pool in which particular things sink or even a cake in which they are the currants. Real things are sharp and knobbly and complicated and different. Everythingism is congenial to our minds because it is the natural philosophy of a totalitarian, mass-producing, conscripted age. That is why we must be perpetually on our guard against it.

And yet . . . and yet . . . It is that *and yet* which I fear more than any positive argument against miracles: that soft, tidal return of your habitual outlook as you close the book and the familiar four walls about you and the familiar noises from the street reassert themselves. Perhaps (if I dare suppose so much) you have been led on at times while you were reading, have felt ancient hopes and fears astir in your heart, have perhaps come almost to the threshold of belief—but now? No. It just won't do. Here is the ordinary, here is the 'real' world, round you again. The dream is ending; as all other similar dreams have always ended. For of course this is not the first time such a thing has happened. More than once in your life before this you have heard a strange story, read some odd book, seen something queer or imagined you have seen it, entertained some wild hope or terror: but always it ended in the same way. And always you wondered how you could, even for a moment, have expected it not to. For that 'real world' when you came back to it is so unanswerable. *Of course* the strange story was false, of course the voice was really subjective, of course the apparent portent was a coincidence. You are ashamed of yourself for having ever thought otherwise: ashamed, relieved, amused, disappointed, and angry all at once. You ought to have known that, as Arnold says, 'Miracles don't happen'.

About this state of mind I have just two things to say. First, that it is precisely one of those counterattacks by Nature which, on my theory, you ought to have anticipated. Your rational thinking has no foothold in your merely natural consciousness except what it wins and maintains by conquest. The moment rational thought ceases, imagination, mental habit, temperament, and the 'spirit of the age' take charge of you again. New thoughts, until they have themselves become habitual, will affect your consciousness as a whole only while you are actually thinking them. Reason has but to nod at his post, and instantly Nature's patrols are infiltrating. Therefore, while counter-arguments against Miracle are to be given full attention (for if I am wrong, then the sooner I am refuted the better not only for you but for me) the mere gravitation of the mind back to its habitual outlook must be discounted. Not only in this enquiry but in every enquiry. That same familiar room, reasserting itself as one closes the book, can make other things *feel* incredible besides miracles. Whether the book has been telling you that the end of civilisation is at hand, that you are kept in your chair by the curvature of space, or even that you are upside down in relation to Australia, it may still seem a little unreal as you yawn and think of going to bed. I have found even a simple truth (e.g. that my hand, this hand now resting on the book, will one day be a skeleton's hand) singularly unconvincing at such a moment. 'Belief-feelings', as Dr Richards calls them, do not follow reason except by long training: they follow Nature, follow the grooves and ruts which already exist in the mind. The firmest theoretical conviction in favour of materialism will not prevent a particular kind of man, under certain conditions, from being afraid of ghosts. The firmest theoretical conviction in favour of miracles will not prevent another kind of man, in other conditions, from *feeling* a heavy, inescapable certainty that no miracle can ever occur. But the feelings of a tired and nervous man, unexpectedly reduced to passing a night in a large empty country house at the end of a journey on which he has been reading a ghost-story, are no evidence that ghosts exist. Your feelings at this moment are no evidence that miracles do not occur.

The second thing is this. You are probably quite right in thinking that you will never see a miracle done: you are probably equally right in thinking that there was a natural explanation of anything in your past life which seemed, at the first glance, to be 'rum' or odd'. God does not shake miracles into Nature at random as if from a pepper-caster. They come on great occasions: they are found at the great ganglions of history—not of political or social history, but of that spiritual history which cannot be fully known by men. If your own life does not happen to be near one of those great ganglions, how should you expect to see one? If we were heroic missionaries, apostles, or martyrs, it would be a different matter. But why you or I? Unless you live near a railway, you will not see trains go past your windows. How likely is it that you or I will be present when a peace-treaty is signed, when a great scientific discovery is made, when a dictator commits suicide? That we should see a miracle is even less likely. Nor, if we understand, shall we be anxious to do so. 'Nothing almost sees miracles but misery'. Miracles and martyrdoms tend to bunch about the same areas of history—areas we have naturally no wish to frequent. Do not, I earnestly advise you, demand an ocular proof unless you are already perfectly certain that it is not forthcoming.

APPENDIX A

ON THE WORDS 'SPIRIT' AND 'SPIRITUAL'

The reader should be warned that the angle from which Man is approached in Chapter IV is quite different from that which would be proper in a devotional or practical treatise on the spiritual life. The kind of analysis which you make of any complex thing depends on the purpose you have in view. Thus in a society the important distinctions, from one point of view, would be those of male and female, children and adults, and the like. From another point of view the important distinctions would be those of rulers and ruled. From a third point of view distinctions of class or occupation might be the most important. All these different analyses might be equally correct, but they would be useful for different purposes. When we are considering Man as evidence for the fact that this spatio-temporal Nature is not the only thing in existence, the important distinction is between that part of Man which belongs to this spatio-temporal Nature and that part which does not: or, if you prefer, between those phenomena of humanity which are rigidly interlocked with all other events in this space and time and those which have a certain independence. These two parts of a man may rightly be called Natural and Supernatural: in calling the second 'Super-Natural' we mean that it is something which invades, or is added to, the great interlocked event in space and time, instead of merely arising from it. On the other hand this 'Supernatural' part is itself a created being—a thing called into existence by the Absolute Being and given by Him a certain character or 'nature'. We could therefore say that while 'supernatural' in relation to *this* Nature (this complex event in space and time) it is, in another sense, 'natural'—i.e. it is a specimen of a class of things which God normally creates after a stable pattern.

There is, however, a sense in which the life of this part can become *absolutely* Supernatural, i.e. not beyond *this* Nature but beyond any and every Nature, in the sense that it can achieve a kind of life which could never have been *given* to any created being in its mere creation. The distinction will, perhaps, become clearer if we consider it in relation not to men but to angels. (It does not matter, here, whether the reader believes in angels or not. I am using them only to make the point clearer.) All angels, both the 'good' ones and the bad or 'fallen' ones which we call devils, are equally

'Super-natural' in relation to *this* spatio-temporal Nature: i.e. they are outside it and have powers and a mode of existence which it could not provide. But the good angels lead a life which is Supernatural in another sense as well. That is to say, they have, of their own free will, offered back to God in love the 'natures' He gave them at their creation. All creatures of course live from God in the sense that He made them and at every moment maintains them in existence. But there is a further and higher kind of 'life from God' which can be given only to a creature who voluntarily surrenders himself to it. This life the good angels have and the bad angels have not: and it is absolutely Supernatural because no creature in any world can have it by the mere fact of being the sort of creature it is.

As with angels, so with us. The rational part of every man is supernatural in the relative sense—the same sense in which *both* angels and devils are supernatural. But if it is, as the theologians say, 'born again', if it surrenders itself back to God in Christ, it will then have a life which is absolutely Supernatural, which is not created at all but begotten, for the creature is then sharing the begotten life of the Second Person of the Deity.

When devotional writers talk of the 'spiritual life'—and often when they talk of the 'supernatural life' or when I myself, in another book, talked of *Zoë*—they mean this *absolutely* Supernatural life which no creature can be given simply by being created but which every rational creature can have by voluntarily surrendering itself to the life of Christ. But much confusion arises from the fact that in many books the words 'Spirit' or 'Spiritual' are also used to mean the *relatively* supernatural element in Man, the element external to *this* Nature which is (so to speak) 'issued' or handed out to him by the mere fact of being created as a Man at all.

It will perhaps be helpful to make a list of the sense in which the words 'spirit', 'spirits' and 'spiritual' are, or have been, used in English.

1. The chemical sense, e.g. 'Spirits evaporate very quickly.'

2. The (now obsolete) medical sense. The older doctors believed in certain extremely fine fluids in the human body which were called 'the spirits'. As medical science this view has long been abandoned, but it is the origin of some expressions we still use; as when we speak of being 'in high spirits' or 'in low spirits' or say that a horse is 'spirited' or that a boy is 'full of animal spirits'.

3. 'Spiritual' is often used to mean simply the opposite of 'bodily' or 'material'. Thus all that is immaterial in man (emotions, passions, memory, etc.) is often called 'spiritual'. It is very important to remember that what is 'spiri-

tual' in this sense is not necessarily good. There is nothing specially fine about the mere fact of immateriality. Immaterial things may, like material things, be good or bad or indifferent.

4. Some people use 'spirit' to mean that relatively supernatural element which is given to every man at his creation—the rational element. This is, I think, the most useful way of employing the word. Here again it is important to realise that what is 'spiritual' is not necessarily good. A Spirit (in this sense) can be either the best or the worst of created things. It is because Man is (in this sense) a spiritual animal that he can become either a son of God or a devil.

5. Finally, Christian writers use 'spirit' and 'spiritual' to mean the life which arises in such rational beings when they voluntarily surrender to Divine grace and become sons of the Heavenly Father in Christ. It is in this sense, and in this sense alone, that the 'spiritual' is always good.

It is idle to complain that words have more than one sense. Language is a living thing and words are bound to throw out new senses as a tree throws out new branches. It is not wholly a disadvantage, since in the act of disentangling these senses we learn a great deal about the things involved which we might otherwise have overlooked. What is disastrous is that any word should change its sense during a discussion without our being aware of the change.

Hence, for the present discussion, it might be useful to give different names to the three things which are meant by the word 'Spirit' in senses three, four, and five. Thus for sense three a good word would be 'soul': and the adjective to go with it would be 'psychological'. For sense four we might keep the words 'spirit' and 'spiritual'. For sense five the best adjective would be 'regenerate', but there is no very suitable noun.[1] And this is perhaps significant: for what we are talking about is not (as *soul* and *spirit* are) a part or element in Man but a redirection and revitalising of all the parts or elements. Thus in one sense there is nothing more in a regenerate man than in an unregenerate man, just as there is nothing more in a man who is walking in the right direction than in one who is walking in the wrong direction. In another sense, however, it might be said that the regenerate man is *totally* different from the unregenerate, for the regenerate life, the Christ that is formed in him, transforms every part of him: in it his spirit, soul and body will all be reborn. Thus if the regenerate life is not a *part* of the man, this is largely because where it arises at all it cannot rest till it becomes the whole man. It is not divided

---

[1] Because the 'spirit' in this sense is identical with the New Man (the Christ formed in each perfected Christian) some Latin theologians call it simply our *Novitas* i.e. our 'newness'.

from any of the parts as they are divided from each other. The life of the 'spirit' (in sense four) is in a sense cut off from the life of the soul: the purely rational and moral man who tries to live entirely by his created spirit finds himself forced to treat the passions and imaginations of his soul as mere enemies to be destroyed or imprisoned. But the regenerate man will find his soul eventually harmonised with his spirit by the life of Christ that is in him. Hence Christians believe in the resurrection of the body, whereas the ancient philosophers regard the body as a mere encumbrance. And this perhaps is a universal law, that the higher you rise the lower you can descend. Man is a tower in which the different floors can hardly be reached from one another but all can be reached from the top floor.

N.B. In the Authorised Version the 'spiritual' man means what I am calling the 'regenerate' man: the 'natural' man means, I think, both what I call the 'spirit man' and the 'soul man'.

APPENDIX B

ON 'SPECIAL PROVIDENCES'

In this book the reader has heard of two classes of events and two only—miracles and natural events. The former are not interlocked with the history of Nature in the backward direction—i.e. in the time before their occurrence. The latter are. Many pious people, however, speak of certain events as being 'providential' or 'special providences' without meaning that they are miraculous. This generally implies a belief that, quite apart from miracles, some events are providential in a sense in which some others are not. Thus some people thought that the weather which enabled us to bring off so much of our army at Dunkirk was 'providential' in some way in which weather as a whole is not providential. The Christian doctrine that some events, though not miracles, are yet answers to prayer, would seem at first to imply this.

I find it very difficult to conceive an intermediate class of events which are neither miraculous nor merely 'ordinary'. Either the weather at Dunkirk was or was not that which the previous physical history of the universe, by its own character, would inevitably produce. If it was, then how is it 'specially' providential? If it was not, then it was a miracle.

It seems to me, therefore, that we must abandon the idea that there is any special class of events (apart from miracles) which can be distinguished as 'specially providential'. Unless we are to abandon the conception of Providence altogether, and with it the belief in efficacious prayer, it follows that all events are equally providential. If God directs the course of events at all then he directs the movement of every atom at every moment; 'not one sparrow falls to the ground' without that direction. The 'naturalness' of natural events does not consist in being somehow outside God's providence. It consists in their being interlocked with one another inside a common space-time in accordance with the fixed pattern of the 'laws'.

In order to get any picture at all of a thing, it is sometimes necessary to begin with a false picture and then correct it. The false picture of Providence (false because it represents God and Nature as being both contained in a common Time) would be as follows. Every event in Nature results from some previous event, not from the laws of Nature. In the long run the first natural event, whatever it was, has dictated every other event. That is, when God at the moment of creation fed the first event into the framework of the 'laws'—first set the ball rolling—He determined the whole history of Nature. Foreseeing every part of that history, He intended every part of it. If He had wished for different weather at Dunkirk He would have made the first event slightly different.

The weather we actually had is therefore in the strictest sense providential; it was decreed, and decreed for a purpose, when the world was made—but no more so (though more interestingly to us) than the precise position at this moment of every atom in the ring of Saturn.

It follows (still retaining our false picture) that every physical event was determined so as to serve a great number of purposes.

Thus God must be supposed in predetermining the weather at Dunkirk to have taken fully into account the effect it would have not only on the destiny of two nations but (what is incomparably more important) on all the individuals involved on both sides, on all animals, vegetables and minerals within range, and finally on every atom in the universe. This may sound excessive, but in reality we are attributing to the Omniscient only an infinitely superior degree of the same kind of skill which a mere human novelist exercises daily in constructing his plot.

Suppose I am writing a novel. I have the following problems on my hands: (1) Old Mr A. has got to be dead before Chapter 15. (2) And he'd better die suddenly because I have to prevent him from altering his will. (3) His daughter (my heroine) has got to be kept out of London for three chapters at least. (4) My hero has somehow got to recover the heroine's good opinion which he lost in Chapter 7. (5) That young prig B. who has to improve before the end of the book, needs a bad moral shock to take the conceit out of him. (6) We haven't decided on B.'s job yet; but the whole development of his character will involve giving him a job and showing him actually at work. How on earth am I to get in all these six things? . . . I have it. What about a railway accident? Old A. can be killed in it, and that settles him. In fact the accident can occur while he is actually going up to London to see his solicitor with the very purpose of getting his will altered. What more natural than that his daughter should run up with him? We'll have her slightly injured in the accident: that'll prevent her reaching London for as many chapters as we need. And the hero can be on the same train. He can behave with great coolness and heroism during the accident—probably he'll rescue the heroine from a burning carriage. That settles my fourth point. And the young prig B.? We'll make him the signalman whose negligence caused the accident. That gives him his moral shock and also links him up with the main plot. In fact, once we have thought of the railway accident, that single event will solve six apparently separate problems.

No doubt this is in some ways an intolerably misleading image: firstly because (except as regards the prig B.) I have been thinking not of the ultimate good of my characters but of the entertainment of my readers: secondly because we are simply ignoring the effect of the railway accident on all the other passengers in that train: and finally because it is I who make B. give the wrong signal. That is, though I pretend that he has free will, he really hasn't. In spite of these objections, however, the example may perhaps suggest how Divine ingenuity could so contrive the physical 'plot' of the universe as to provide a 'providential' answer to the needs of innumerable creatures.

But some of these creatures have free will. It is at this point that we must begin to correct the admittedly false picture of Providence which we have hitherto been using. That picture, you will remember, was false because it represented God and Nature as inhabiting a common Time. But it is probable that Nature is not really in Time and almost certain that God is not. Time is probably (like perspective) the mode of our perception. There is therefore in reality no question of God's at one point in time (the moment of creation) adapting the material history of the universe in advance to free acts which you or I are to perform at a later point in Time. To Him all the physical events and all the human acts are present in an eternal Now. The liberation of finite wills and the creation of the whole material history of the universe (related to the acts of those wills in all the necessary complexity) is to Him a single operation. In this sense God did not create the universe long ago but creates it at this minute—at every minute.

Suppose I find a piece of paper on which a black wavy line is already drawn, I can now sit down and draw other lines (say in red) so shaped as to combine with the black line into a pattern. Let us now suppose that the original black line is conscious. But it is not conscious along the whole length at once—only on each point on that length in turn.

Its consciousness in fact is travelling along that line from left to right retaining point A only as a memory when it reaches B and unable until it has left B to become conscious of C. Let us also give this black line free will. It chooses the direction it goes in. The particular wavy shape of it is the shape it wills to have. But whereas it is aware of its own chosen shape only moment by moment and does not know at point D which way it will decide to turn at point F, I can see its shape as a whole and all at once. At every moment it will find my red lines waiting for it and adapted to it. Of course: because I, in composing the total red-and-black design have the whole course of the black line in view and take it into account. It is a matter not of impossibility but merely of designer's skill for me to devise red lines which at every point have a right relation not only to the black line but to one another so as to fill the whole paper with a satisfactory design.

In this model the black line represents a creature with free will, the red lines represent material events, and I represent God. The model would of course be more accurate if I were making the paper as well as the pattern and if there were hundreds of millions of black lines instead of one—but for the sake of simplicity we must keep it as it is.[1]

It will be seen that if the black line addressed prayers to me I might (if I chose) grant them. It prays that when it reaches point N it may find the red lines arranged around it in a certain shape. That shape may by the laws of design require to be balanced by other arrangements of red lines on quite different parts of the paper—some at the top or bottom so far away from the black line that it knows nothing about them: some so far to the left that they come

[1] Admittedly all I have done is to turn the tables by making human volitions the constant and physical destiny the variable. This is as false as the opposite view; the point is that it is no falser. A subtler image of creation and freedom (or rather, creation of the free and the unfree in a single timeless act) would be the *almost* simultaneous mutual adaptation in the movement of two expert dancing partners.

before the beginning of the black line, some so far to the right that they come after its end. (The black line would call these parts of the paper, 'The time before I was born,' and, 'The time after I'm dead.') But these other parts of the pattern demanded by that red shape which Black Line wants at N, do not prevent my granting its prayer. For his whole course has been visible to me from the moment I looked at the paper and his requirements at point N are among the things I took into account in deciding the total pattern.

Most of our prayers if fully analysed, ask either for a miracle or for events whose foundation will have to have been laid before I was born, indeed, laid when the universe began. But then to God (though not to me) I and the prayer I make in 1945 were just as much present at the creation of the world as they are now and will be a million years hence. God's creative act is timeless and timelessly adapted to the 'free' elements within it: but this timeless adaptation meets our consciousness as a sequence and prayer and answer.

Two corollaries follow:

1. People often ask whether a given event (not a miracle) was really an answer to prayer or not. I think that if they analyse their thought they will find they are asking, 'Did God bring it about for a special purpose or would it have happened anyway as part of the natural course of events?' But this (like the old question, 'Have you left off beating

your wife?') makes either answer impossible. In the play, *Hamlet,* Ophelia climbs out on a branch overhanging a river: the branch breaks, she falls in and drowns. What would you reply if anyone asked, 'Did Ophelia die because Shakespeare for poetic reasons wanted her to die at that moment—or because the branch broke?' I think one would have to say, 'For both reasons.' Every event in the play happens as a result of other events in the play, but also every event happens because the poet wants it to happen. All events in the play are Shakespearian events; similarly all events in the real world are providential events. All events in the play, however, come about (or ought to come about) by the dramatic logic of events. Similarly all events in the real world (except miracles) come about by natural causes. 'Providence' and Natural causation are not alternatives; both determine every event because both are one.

2. When we are praying about the result, say, of a battle or a medical consultation the thought will often cross our minds that (if only we knew it) the event is already decided one way or the other. I believe this to be no good reason for ceasing our prayers. The event certainly has been decided—in a sense it was decided 'before all worlds'. But one of the things taken into account in deciding it, and therefore one of the things that really cause it to happen, may be this very prayer that we are now offering. Thus,

shocking as it may sound, I conclude that we can at noon become part causes of an event occurring at ten a.m. (Some scientists would find this easier than popular thought does.) The imagination will, no doubt, try to play all sorts of tricks on us at this point. It will ask, 'Then if I stop praying can God go back and alter what has already happened?' No. The event has already happened and one of its causes has been the fact that you are asking such questions instead of praying. It will ask, 'Then if I begin to pray can God go back and alter what has already happened?' No. The event has already happened and one of its causes is your present prayer. Thus something does really depend on my choice. My free act contributes to the cosmic shape. That contribution is made in eternity or 'before all worlds'; but my consciousness of contributing reaches me at a particular point in the time-series.

The following question may be asked: If we can reasonably pray for an event which must in fact have happened or failed to happen several hours ago, why can we not pray for an event which we know *not* to have happened? e.g. pray for the safety of someone who, as we know, was killed yesterday. What makes the difference is precisely our knowledge. The known event states God's will. It is psychologically impossible to pray for what we know to be unobtainable; and if it were possible the prayer would sin against the duty of submission to God's known will.

One more consequence remains to be drawn. It is never possible to prove empirically that a given, nonmiraculous event was or was not an answer to prayer. Since it was non-miraculous the sceptic can always point to its natural causes and say, 'Because of these it would have happened anyway,' and the believer can always reply, 'But because these were only links in a chain of events, hanging on other links, and the whole chain hanging upon God's will, they may have occurred because someone prayed.' The efficacy of prayer, therefore, cannot be either asserted or denied without an exercise of the will—the will choosing or rejecting faith in the light of a whole philosophy. Experimental evidence there can be none on either side. In the sequence M.N.O. event N, unless it is a miracle, is always caused by M and causes O; but the real question is whether the total series (say A–Z) does or does not originate in a will that can take human prayers into account.

This impossibility of empirical proof is a spiritual necessity. A man who knew empirically that an event had been caused by his prayer would feel like a magician. His head would turn and his heart would be corrupted. The Christian is not to ask whether this or that event happened because of a prayer. He is rather to believe that all events without exception are *answers* to prayer in the sense that whether they are grantings or refusals the prayers of all concerned and their needs have all been taken into account. All prayers are heard, though not all prayers are granted. We must not picture destiny as a film unrolling for the most part on its own, but in which our prayers are sometimes allowed to insert additional items. On the contrary; what the film displays to us as it unrolls already contains the results of our prayers and of all our other acts. There is no question *whether* an event has happened because of your prayer. When the event you prayed for occurs your prayer has always contributed to it. When the opposite event occurs your prayer has never been ignored; it has been considered and refused, for your ultimate good and the good of the whole universe. (For example, because it is better for you and for everyone else in the long run that other people, including wicked ones, should exercise free will than that you should be protected from cruelty or treachery by turning the human race into automata.) But this is, and must remain, a matter of faith. You will, I think, only deceive yourself by trying to find special evidence for it in some cases more than in others.